GAMES & ACTIVITIES/with base ten blocks-I

by

Rebecca S. Nelson

Cuisenaire Company of America, Inc.
12 Church St., Box D
New Rochelle, N.Y. 10802

Cover Illustration by **Joshua M. Berger**

Cuisenaire Company of America, Inc.

12 Church Street, Box D, New Rochelle, New York 10802

Printed in the United States of America
ISBN 0-914040-57-X

TABLE OF CONTENTS

INTRODUCTION

This book is mostly about place value and trades. It has been my experience, as I work with children in a clinical situation, that the ability to perform two and three-digit operations depends mostly upon an understanding of place value. In this book, I will use the word *trade* rather than words such as *regroup, rename, borrow* or *carry*. There are several reasons for doing this. First of all trade is a word that children understand because they can do it. Secondly, they have one word to learn, rather than several words which essentially mean the same thing. Finally, does anyone ever pay back what they borrowed? Where do you carry to? And how does it magically (to most children) change from ten to one just because it was carried?

This book will attempt to emphasize those areas which textbooks leave out or can't address well because of space limitations or limitations of the medium. Consider as an example the idea of "renaming" 13 ones as 1 ten 3 ones. (This is usually presented just before two digit addition with trades.) A sample item might look similar to this:

13 ones = _______ tens _______ ones

My questioning of children reveals that they don't see or think 13 ones. They look at one box and 3 left and fill in the correct answer.

This book will also introduce some games. These games will have three characteristics in common. First, all games will actually use the base ten blocks, and pictures or diagrams of base ten blocks. This is designed so that children can figure out the answer while playing the game and thus the student who needs the most practice is not shut out or penalized. Secondly, all games have an element of chance (roll of the dice, spin of a spinner, draw of the card) which gives all children, regardless of skill, a chance to win. This element provides motivation for the less skilled and promotes cooperation rather than hostility among the children playing. Thirdly, all games are designed for small group play. This characteristic provides an opportunity for classroom individualization. Several copies of one game can be constructed and used with children grouped by ability. Different games can be assigned to different children depending on need, or games can be used at a learning center. Although games used every day could become boring, used wisely, they do provide a motivational alternative to practice or drill sheets.

CONSTRUCTING THE GAMES

Gameboards, playing boards, and cards in the appendix of this book are designed to be reproduced. Frequently a modification of the same board or cards will be used for several games. Gameboards may be mounted on file folders or poster board for extended wear and convenience of storage. Running gameboards off on a ditto machine on construction paper is sometimes preferred because they last surprisingly well and need not be stored. Game cards may be mounted on 3 x 5 index cards. If different colored cards are used for the different games (especially those of about the same level), it will be easier to keep the decks sorted and organized. Spinners are easier for children to handle if mounted on paper plates. The spinning arrow is best constructed from poster board or other light weight cardboard. Punching the hole with a paper punch allows the arrow to spin more freely. Also, the paper fastener used to hold the arrow to the spinner should be loose. Taping it on the back side of the plate will help hold the fastener so the arrow may spin easily. Place value mats M4 and M16 are provided for your convenience. If you have mats in your base ten kits, they may be used instead.

READINESS FOR PLACE VALUE

Before children are ready to think about place value concepts, they need to be able to match sets and numerals for numbers through ten; compare greater than, less than or equal to; and, read and write numerals up to and including 10. Books such as *DEVELOPING NUMBER CONCEPTS USING UNIFIX CUBES* by Kathy Richardson and *WORKJOBS* by Mary Baratta-Lorton along with children's games such as *High-Ho-Cherry-O* and *Run Mousey Run*, and selected text activities can provide this entry behavior for most children.

MAKING TENS

Before children can think "tens and ones" or use the base ten rods, they need to make rows of ten. For this activity you will need small paper bags marked with letters A, B, C... Each child or pair of children in the group should have a bag. Place between ten and forty-five small cubes in each bag. Have the children pour the cubes on their desks and make as many rows of ten as possible. Walk around and check. Ask individual children: "How many tens did you make?" Have the children trade bags and repeat. As independent seat work, have each child do the same activity, this time recording on paper the letter of the bag and how many tens.

NAME ______________________ MATH

BAG	HOW MANY TENS?
1. ______	______
2. ______	______
3. ______	______
4. ______	______
5. ______	______

making tens • 2-5 players

You Need: Gameboard M1-M2, "top" spinner M3, paper bags A through L with between 10 and 45 base ten small cubes in each, small marker for each player.

How To Play: Each player spins. The player who spins the letter earliest in the alphabet starts. At each turn the player spins the spinner, gets the bag indicated, pours out the cubes and makes all the rows of ten possible. He/she then moves ahead one place on the gameboard for each row of ten that was made. The first player to reach HOME wins the game.

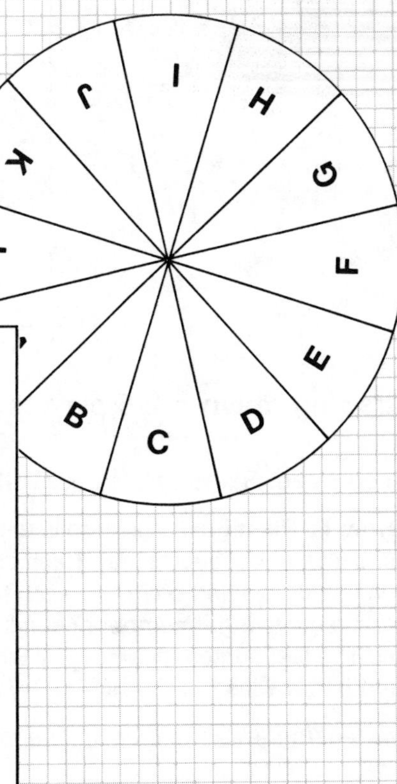

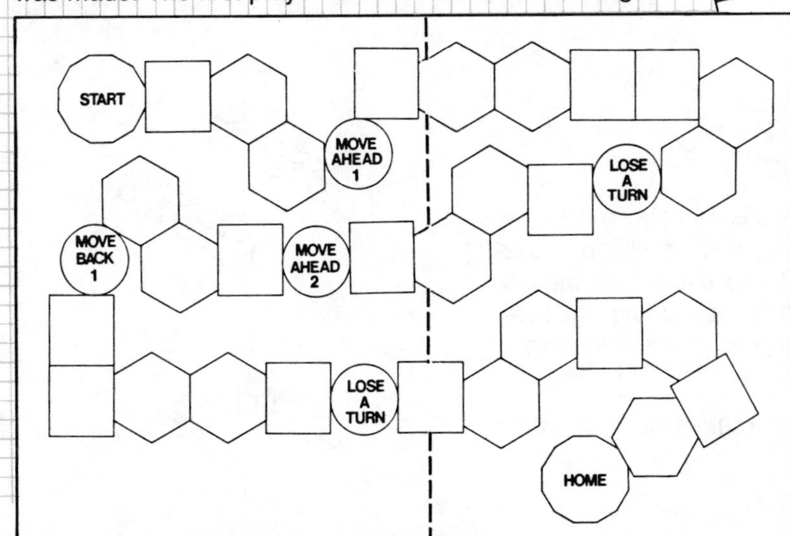

TRADES

Having made their rows of tens, the children are ready to trade.

trading up • 2-5 players

You Need: Base ten rods and small cubes, playing board M4 for each player, die.

How To Play: Each player rolls the die. The player with the smallest number begins play. Players take turns rolling the die and collecting small cubes. When a player has collected ten cubes, he/she may trade them for a rod. The first player to collect five rods wins the game.

Extension: *Use two dice and play for larger amounts such as nine rods instead of five.*

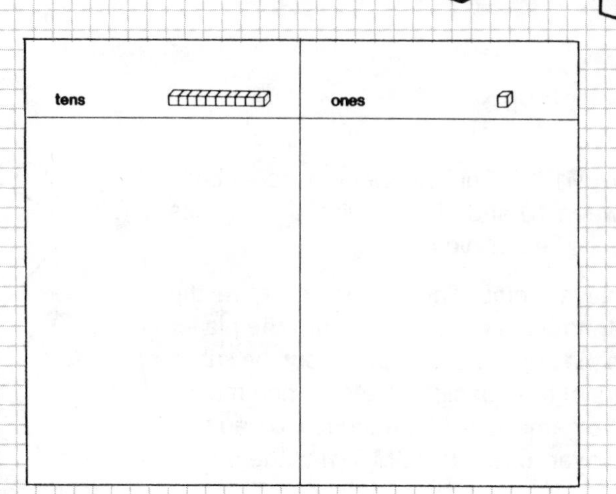

trading down • 2-5 players

You Need: Same as TRADING UP

How To Play: Each player rolls the die. The player with the largest number begins play. Each player starts with four rods on his/her playing board. At each turn a player rolls the die and removes that number of small cubes, trading as needed. The first person to remove all the wood from his/her playing board wins.

Extension: *Use two dice and begin with nine rods on each playing board.*

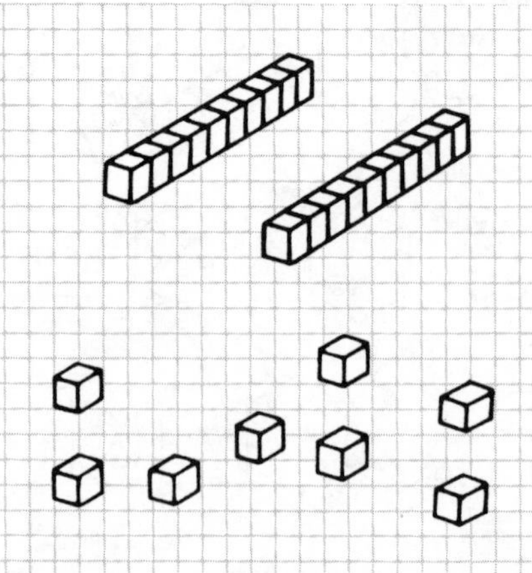

COUNTING BY TENS

Once the children are familiar with the base ten rods, they are ready to count by tens through ninety. Although this is traditionally done later in the curriculum, counting by tens can actually help children learn to count by ones to ninety-nine since they will be less likely to get hung up on what comes after a number such as forty-nine, fifty-nine, or sixty-nine. For this activity, provide each child or pair of children with nine base ten rods and playing board M4. You may wish to make an overhead transparency of the playing board to work along with the children. Have the children place the rods on their board one at a time and say "one ten", "two tens", "three tens" and so on. As they become familiar with this, shorten the word "tens" to "t" and say "one-t", "two-t", "three-t", "four-t", up to "nine-t". This will prepare children to read two-digit numbers such as 46 as "four-t-six". After the children accomplish counting by tens in sequence, they are ready to jump around. For example, have the children show "four-t" with rods. Ask: "When counting by tens,what comes next after four-t?" Have the children show it and say it.

TENS AND ONES

The children should now be ready to begin tens and ones which will eventually lead to meaningful counting as opposed to rote counting. Provide the children with playing board M4, rods, and small cubes. Have the children place 4 rods and 3 cubes on the board. Ask: "How many tens?" Say: "So we have four-t. How many ones? So we have four-t three." Have the children say the words as you point to the rods and small cubes. Continue using other two-digit numbers. Since the teen numbers are the most difficult, introduce them only after the children seem comfortable, and in the early stage, call them one-t-one, one-t-two, one-t-three, up to one-t-nine. As the children show mastery of the verbal, introduce the written symbol as 4^t6, 3^t7, and 5^t4.

t • 2-4 players

You Need: Base ten rods and small cubes, gameboard M1-M2, small marker for each player, deck of cards made from M5 and M6.

To prepare cards, write the "t" numeral on the back of each card.

How To Play: The tallest player in the group begins play. Shuffle the cards and place the numeral side up. On a turn, a player draws the top card, reads the numeral aloud and shows it with rods and small cubes. He/she then checks by looking at the back of the card, correcting if necessary. The player moves one space forward on the gameboard for each rod. The first person to reach HOME wins.

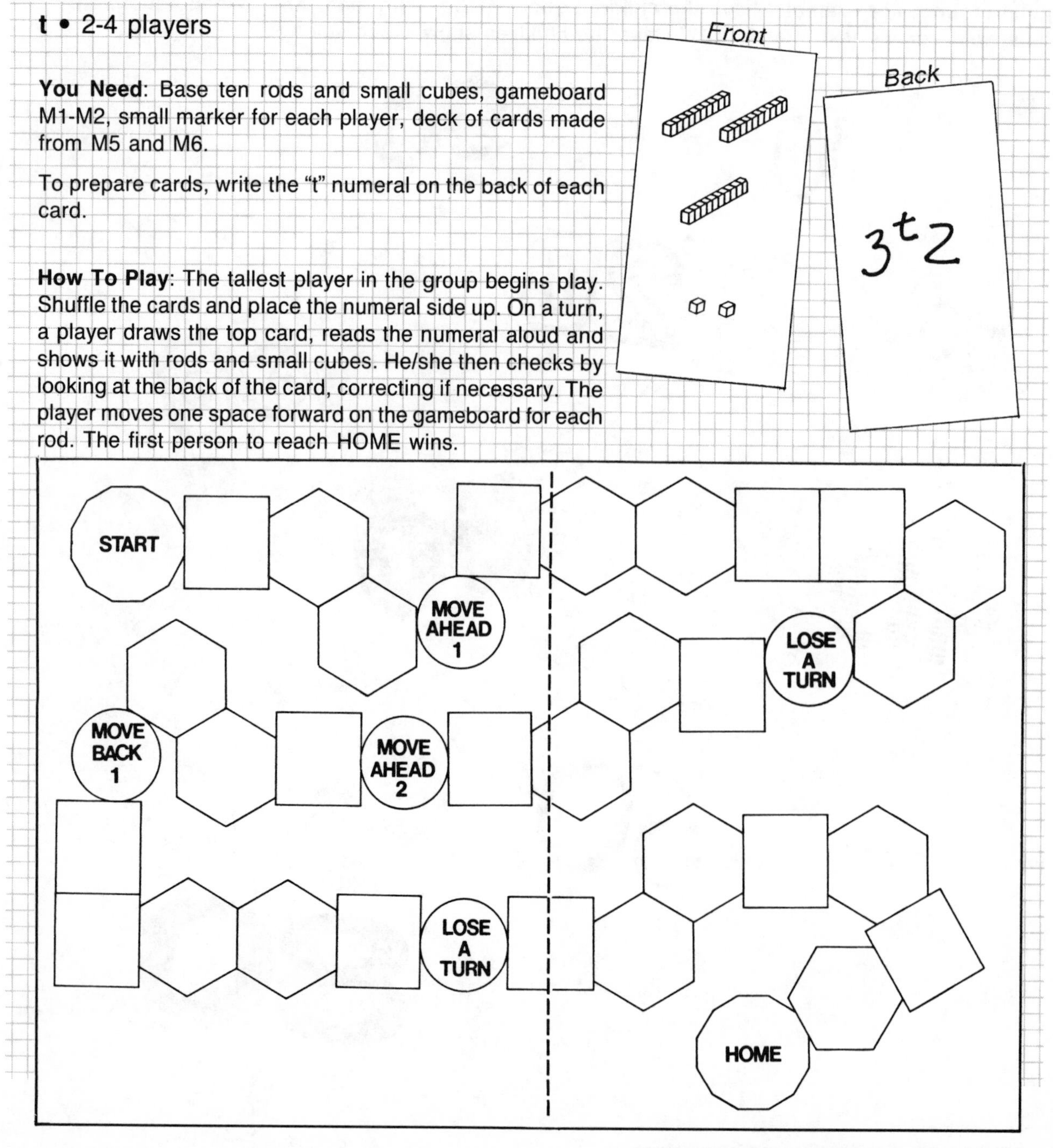

STANDARD TWO-DIGIT NUMERALS

After the children gain confidence using the "t" numeral, introduce the standard numeral. Have the children practice reading standard numerals and showing them with rods and small cubes.

show it • 2-5 players

You Need: Base ten rods and small cubes, playing board M7 for each player, "bottom" spinner M3.

How To Play: Each player spins. The player who spins the largest amount of wood begins. Players take turns spinning and collecting the amount of wood shown. When enough wood has been collected, a player may claim a square on his/her playing board by placing the wood to show the number of that square. The first person to claim three consecutive squares in a row, column, or diagonal on his/her playing board wins the game. Trades may be made during any turn.

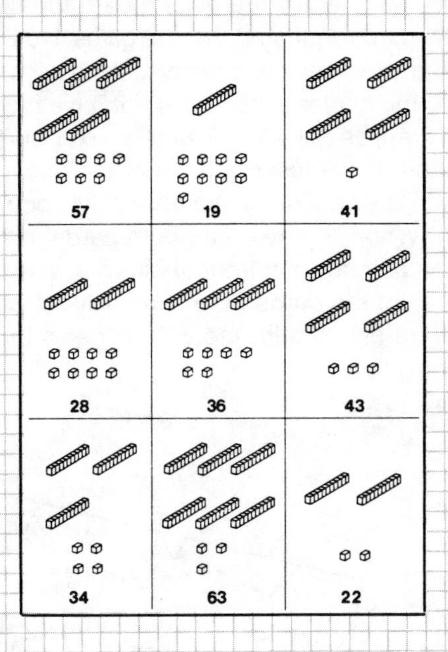

Extension 1: *Show only the numerals on the playing board and play as before.*

Extension 2: *Two players only as in tic-tac-toe. As a player collects enough wood, he/she claims the square with a marker and gives up that amount of wood. The first person to have three in a consecutive row, column, or diagonal wins the game.*

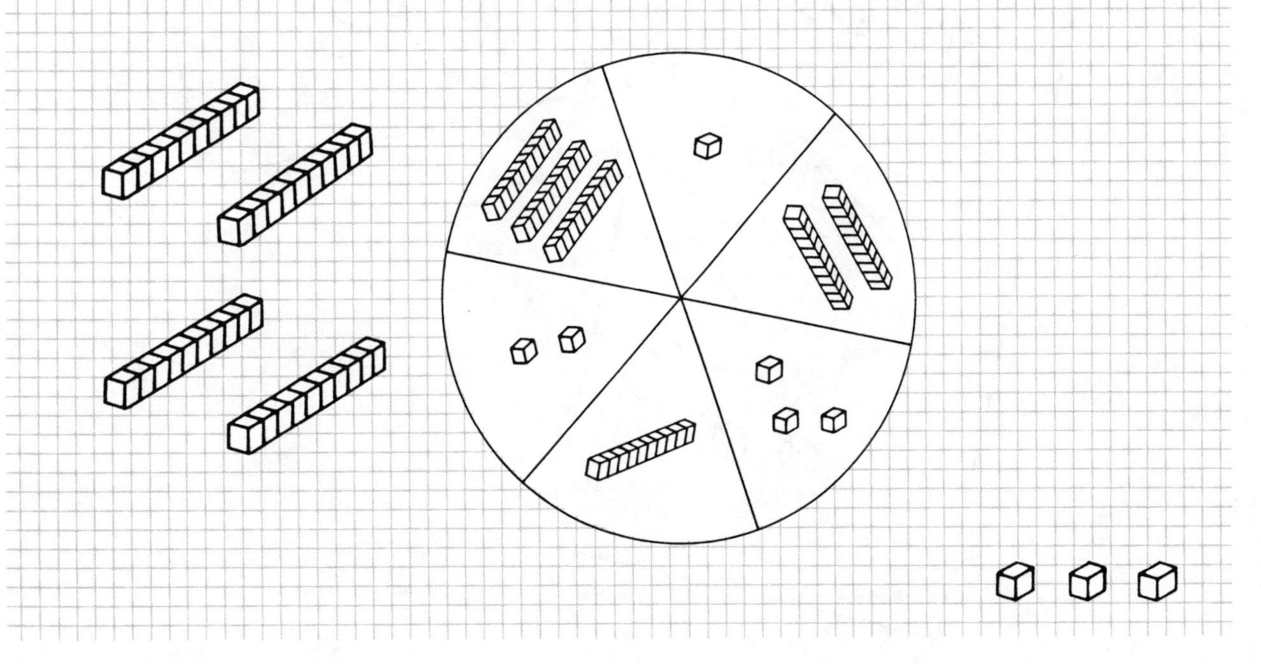

draw • 2-4 players

You Need: Deck of 40 cards, containing 20 picture cards from M5 and M6 and 20 numeral cards made to match the picture cards.

How To Play: The dealer is the shortest person in the group. The dealer shuffles the deck, deals 5 cards to each player and places the other cards face down in the area. Play begins by the dealer asking any other player for one card (Example: "May I have the picture card for 46?"). If the player has the card requested, he/she gives it to the requester and the requester may continue by asking any player for another card. If the player asked does not have the card requested, he/she replies "DRAW". The requester must then draw one card from the unused portion of the deck and his/her turn is over. The person to the left of the requester then takes a turn. When a player has both cards from a pair, he/she lays the pair down in front of him/her. The first player to lay down all of the cards from his/her hand wins. If this group wishes to play again, the winner deals the next game.

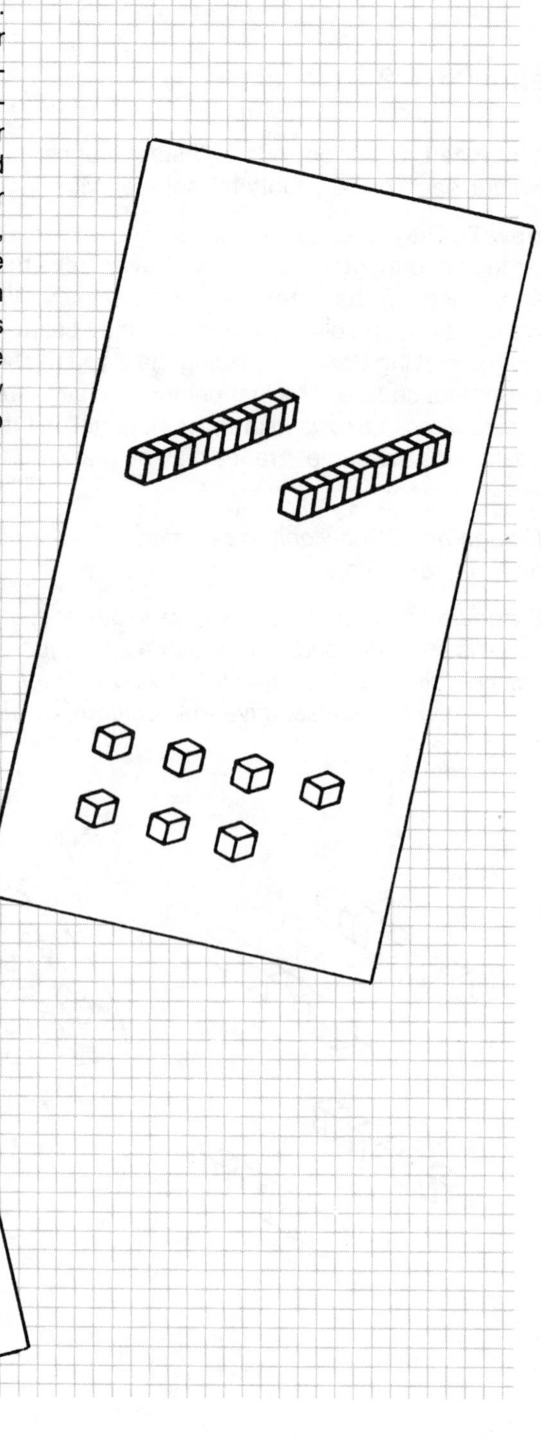

debt • 2-4 players

You Need: Gameboard M1-M2 modified as shown below, die, small marker for each player, base ten rods and small cubes, DEBT cards.

Make ten to fifteen DEBT cards with sayings such as "move backward 2 spaces", "roll again and move forward", "bonus, get 24 in rods and cubes",or "lose a turn".

How To Play: Each player begins with 2 rods and 10 small cubes. Each player rolls the die. The player who rolls the highest number begins play. On a turn, a player rolls the die and moves forward that number of spaces. The player then collects rods and small cubes from the "bank" or pays rods and small cubes to the "bank" as stated. Rods may be traded for small cubes or small cubes for rods to make "correct change" at any time. If a player lands on a PAY square and cannot pay, he/she draws a DEBT card and follows the directions. The first person to reach HOME wins the game.

***Extension**: Use numerals only on the gameboard and play as before.*

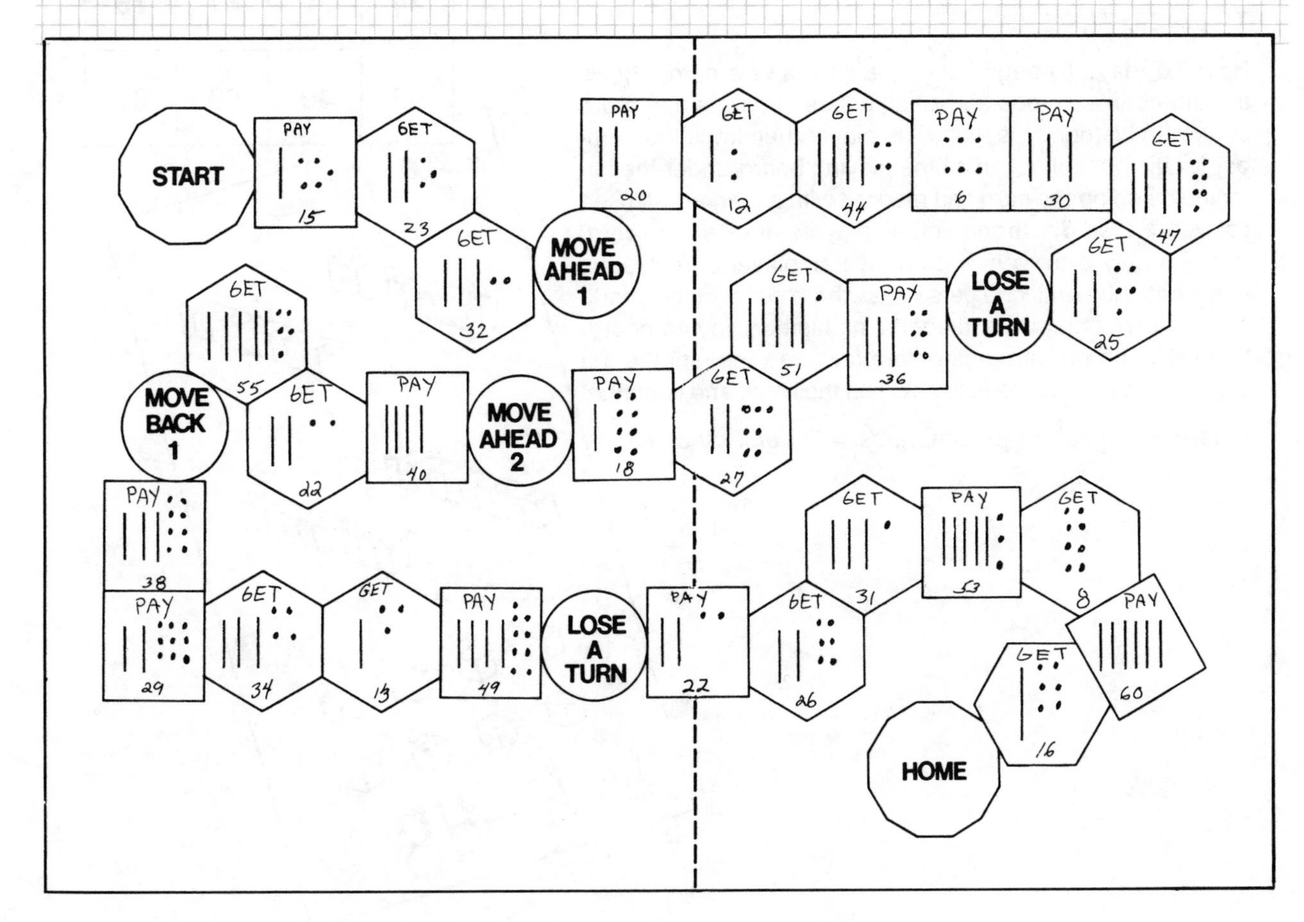

ONE MORE

One more is a very important concept for meaningful counting. Provide rods and small cubes for each student. Say or write a two-digit number. Have the students show it with rods and small cubes, add one more cube and say this new number. Start with numbers such as 23 or 46 which do not require trades and progress to numbers such as 29 and 59 which do require trades.

next • 2-4 players

You Need: Base ten rods and small cubes; picture cards M5 and M6; playing board M8, prepared as shown, one for each player; 40 to 60 markers.

To prepare the playing board, cut a 4 by 6 portion of M8 and fill in as shown.

N	E	X	T
18	21	42	66
24	28	47	70
33	31	52	75
34	39	57	76
43	49	60	85

To prepare the deck of cards, write a letter and a number for each card.

N - 17, 23, 32, 33, 42
E - 20, 27, 30, 38, 48
X - 41, 46, 51, 56, 59
T - 65, 69, 74, 75, 84

How To Play: To begin, each player draws a card, shows that number with rods and small cubes and puts one more cube with his/her collection. The player then finds the number for his/her collection on the playing board under the letter indicated on the card and marks it with a marker. All used cards are then discarded and all players draw a new card. (No one may draw a new card until all players are ready.) Play continues until one person has three consecutive markers in a row, column, or diagonal on his/her playing board. If players run out of cards before anyone wins, shuffle the discarded cards, place face down on the stack and continue

***Extension**: Replace picture cards with numeral cards.*

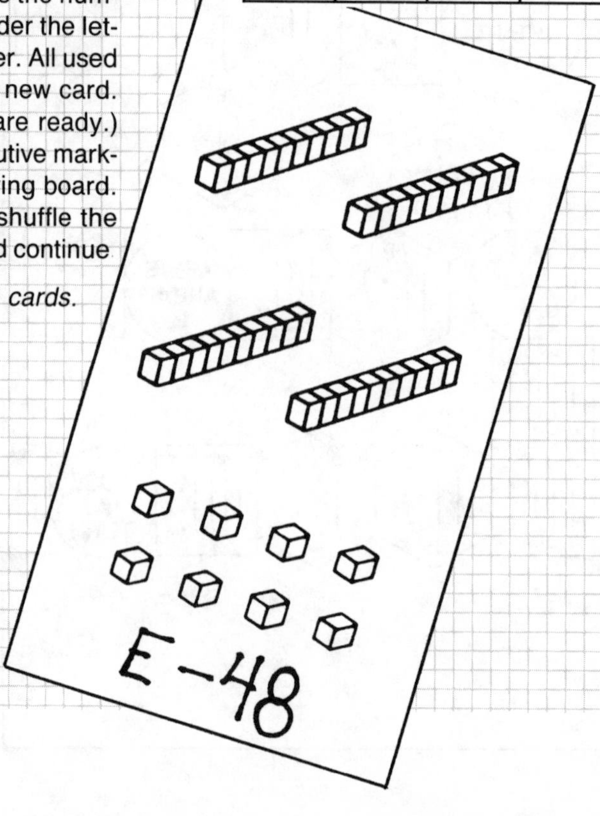

E-48

one more • 4-5 players

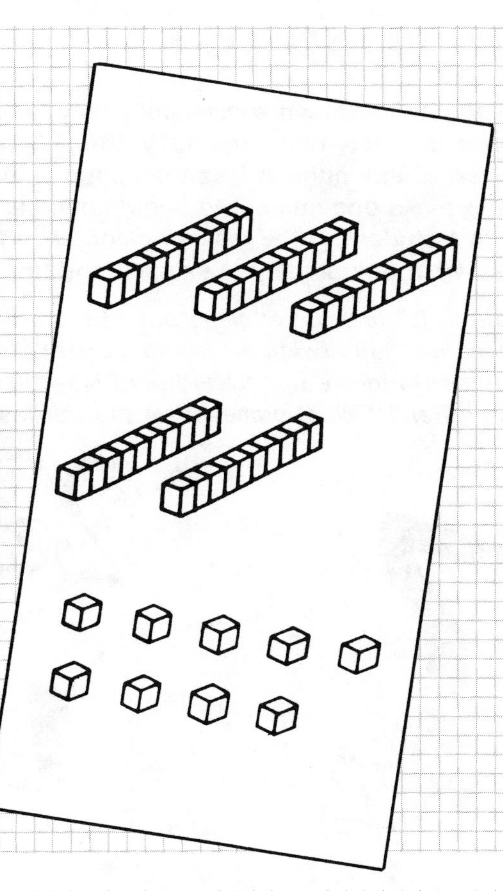

You Need: Deck of 40 cards, containing 20 picture cards M5 and M6 and 20 numeral cards made by using the numbers on the NEXT playing board.

How To Play: Shuffle the numeral cards and lay face down on the table. Each player draws one card. The dealer is the player drawing the highest number. The dealer shuffles the picture cards and deals clockwise, one to each person, until they are gone. The next player starts the game, by shuffling the numeral cards and placing them face down on the table. On a turn, a player draws one picture card from the player on his/her right and one numeral card from the deck. If the numeral card is one more than any picture card in his/her hand, the player lays down the "pair" and draws another numeral card. If not, the player discards the numeral card randomly into the pile and the next player on the left takes his/her turn. The winner is the first person to run out of picture cards.

mine • A caller and 2, 4 or 5 players

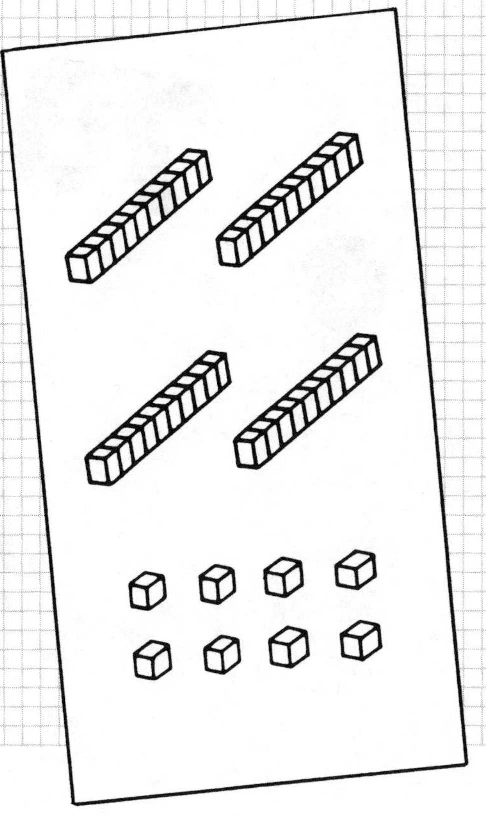

You Need: Deck of 40 cards used in ONE MORE.

How To Play: The caller shuffles and deals the numeral cards to each player until they are gone. He/she shuffles the picture cards and places them face down in front of him/her. One at a time, the caller turns over a picture card on the table (visible to all players) and calls out the number. The person who has the number that is one more calls "MINE", collects both cards and lays the pair on the table in front of him/her. The first person to get rid of all the numeral cards in his/her hand wins the game.

***Extension**: The caller calls the number but does not show the picture to the players.*

COUNTING TO 99

Once the children can successfully play NEXT, ONE MORE, and MINE, they are ready to count by ones to ninety-nine. Hopefully, the children will not necessarily need to begin with one but can start at any number less than ninety-nine and continue. Have the children sit in a circle. The teacher picks one student to begin counting by ones slowly. The teacher claps his/her hands to stop this student. If the teacher claps once the next student picks up the count. If the teacher claps twice, one student is skipped and the second student picks up the count and so on.

Variation: If the teacher claps once the next student picks up the count but skips one number. If the teacher claps twice the same student picks up the count but skips two numbers. Introduce the students to the hundreds board M9-M10 and look for patterns. Ask questions such as "What comes after 3? What comes after the thirties?"

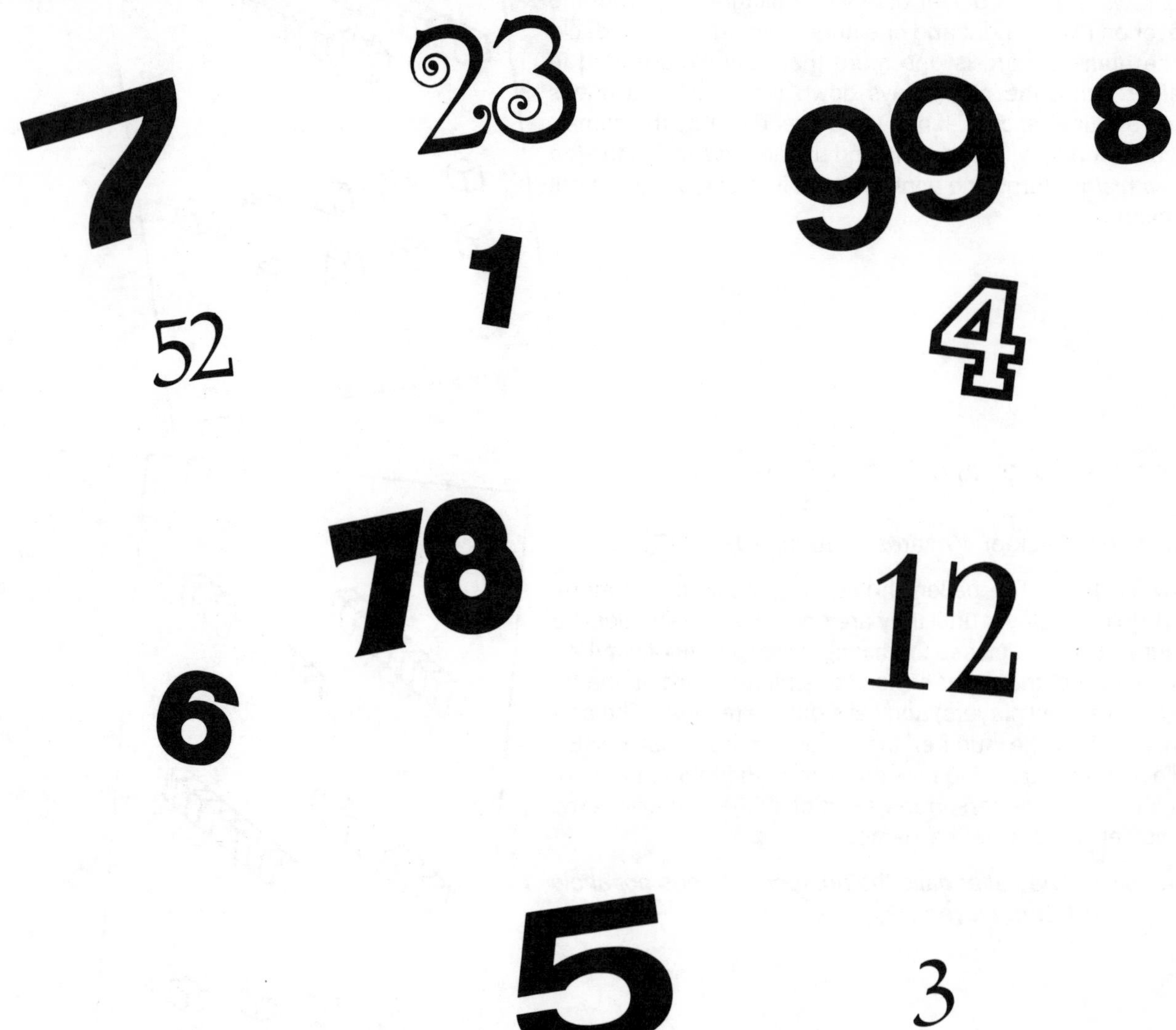

COMPARING TWO-DIGIT NUMBERS

The children already have experience comparing numbers less than ten. For example, 3 is less than 8. Have the children work in pairs. Provide rods and small cubes. Write two numbers such as 24 and 37 on the board. Have one student in the pair show each number. Ask: "Which has more wood? How can we tell easily? How shall we write this?" As the students become confident, move to pairs of numbers with more tens but fewer ones such as 52 and 37. Continue showing the wood and recording.

rod war • 2 players

You Need: Picture cards M5 and M6, reproduced three times for a total of 60 cards, with the number shown written on each; base ten rods and small cubes.

How To Play: Shuffle the cards and give thirty to each player. A player begins with all of his/her cards face down in front of him/her. At the same time, each player turns over his/her top card. The player showing the larger number wins both cards. If both players show the same number they are at "War" and turn over the next card. The player with the larger number now wins all four cards. The winner is the person who collects the most cards. All questions are resolved by showing the numbers with rods and small cubes.

***Extension**: Use number cards rather than picture cards and play as before.*

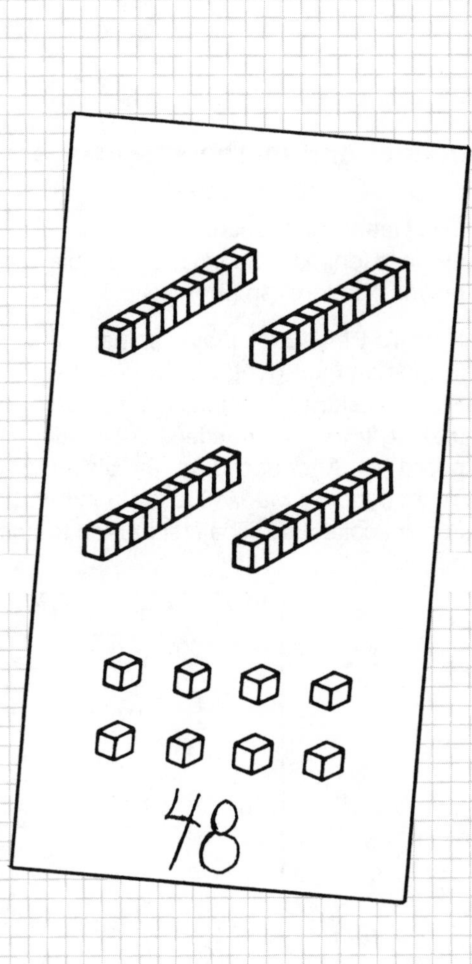

four in a row • 2 players

You Need: 20 picture cards M5 and M6 with the number shown written on each.

How To Play: Each player draws one card. The person having the larger number is the dealer. The dealer shuffles the deck and deals four cards in a row, face up to each player. These cards may not be rearranged. The remainder of the deck is then placed face down on the table with the top card turned over to start a discard pile. On a turn, a player draws the top card from the deck or the top card from the discard pile. The player may discard this card or trade it for one of his/her four cards and then discard. The first person to get four cards ordered from least to greatest wins.

***Extension**: Replace picture cards with numeral cards only and play as before.*

TEN MORE THAN ANY TWO-DIGIT NUMBER

Children frequently learn to say multiples of ten in order (ten, twenty, thirty, and so on) but do not know ten more than 46 or ten more than 37 or 52. For this activity, again provide base ten rods, small cubes and playing mat M4. Have the children show a two-digit number such as 52. Ask: "How can we show one more than this number? How can we show ten more than this number?" Have the children show it. "What number is ten more than 52?" Continue until the children feel comfortable, then record as you work. Look for patterns in the record.

Number	Ten More
52	62
47	57
80	90

collect and march • 2-5 players

You Need: Gameboard M9-M10, small marker for each player, base ten rods and small cubes, playing board M4 for each player, "bottom" spinner M3.

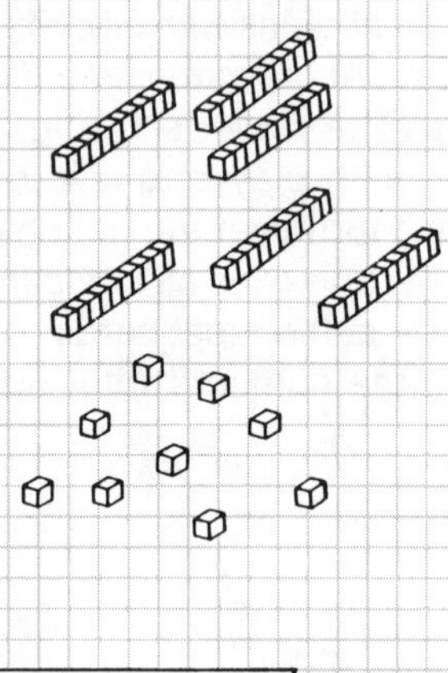

How To Play: Each player spins. The person who spins the largest amount goes first. All markers are placed on start. Players take turn spinning and adding to their collection, making trades as needed. Collections are kept on playing board M4. After each turn, the player moves his/her marker on gameboard M9-M10 to the number that matches his/her current collection. The first person to go past END wins the game.

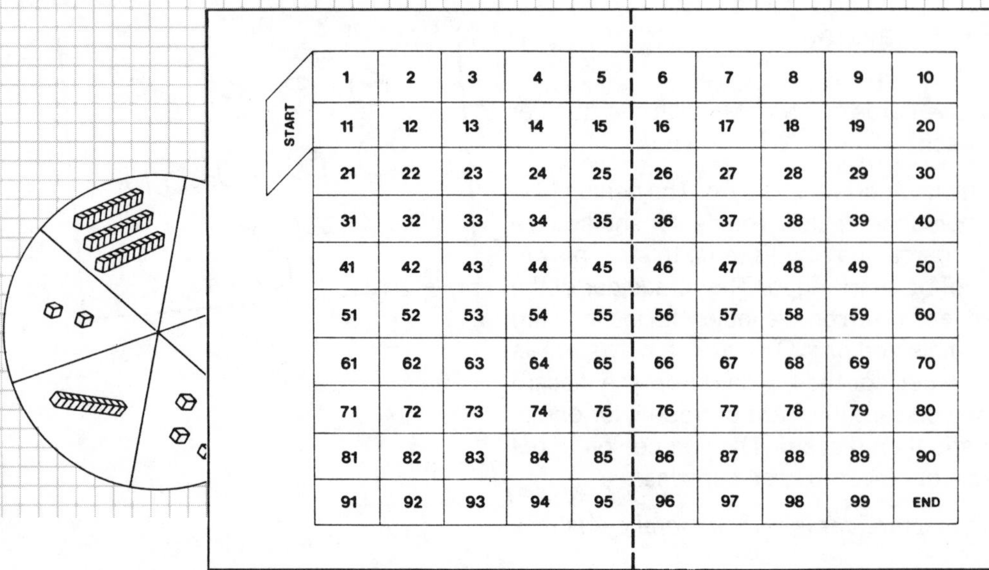

***Extension**: Have the children look at hundreds board M9-M10. Ask questions such as "If you are on 34 and get ten more where will you be? What is an easy way to show ten more? Twenty more?" When the children are comfortable, allow them to play without rods and cubes and collections.*

ten more • 3-4 players

You Need: Deck of 21 cards containing 20 picture cards M5 and M6 and a TEN card shown below.

How To Play: Each person draws a picture card. The person drawing the smallest amount begins as dealer. All cards are shuffled and dealt one at a time clockwise until all cards are distributed. Play begins with the next person after the deck runs out.

All players look at their cards and make as many pairs as possible. A pair is two cards where the number on one card is ten more than the number shown on the other. Pairs are laid on the table in front of the player. Play begins by the player turning to the person on his/her right and drawing a card. If that card completes a pair with one in the player's hand, the pair is laid on the table. If not, the player keeps the card. The next player on the left now takes his/her turn. The first person to get rid of all cards in his/her hand wins. The person left holding the TEN card at the end of the game loses and becomes the dealer for the next game if played.

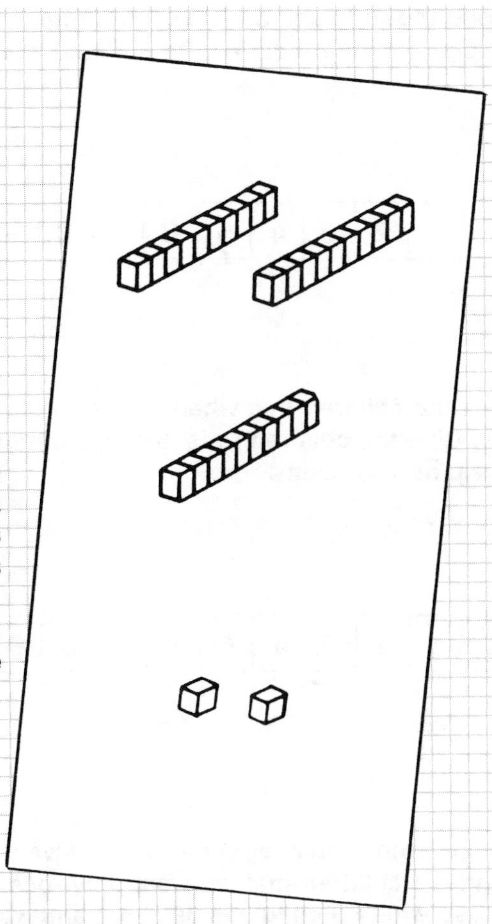

ROUNDING TO THE NEAREST TEN

Here is a concept that many children do not comprehend. Many learn a "rule" with little or no understanding, while others simply become confused and/or lost. Since number lines are very abstract models for most young children, explanations using number lines are frequently of little value.

For this activity, we will use the combination of a number strip (slightly less abstract number line) and the rods and small cubes. Prepare a number strip by cutting gameboard M9-M10 into strips and hooking them end to end and adding a zero as shown below:

Tell the children that when we "round to tens" the tens or rods are our only possible answers. Give the children rods and count together by tens. Color in multiples of ten on the number strip as you count.

Begin with a number such as 39. Have the children show this number with rods and cubes. Remind the children that we want only rods because we are rounding to tens. Liken the situation to a case where a child has 39 cents and wants only dimes. "Would your Mother or Dad give in and trade you a dime for 9 pennies? Notice on the number strip how close 39 is to 40." Tell the children that since it is so close, the *nice rod trader* will let you have another rod for your 9 cubes. Continue with other numbers deciding for each number if it is close enough for the *rod trader* to trade your cubes for another rod, or if it is so far away that the *rod trader* will take the cubes you have. For each example, find the number on the number strip and check how far away it is. Be sure to include rounded numbers such as 40 or 70. This is a good reminder that we want all rods one way or the other. As the children become good at this practice, make a record of what happened.

Took away cubes: 43, 21, 83, 54, 72

Traded for another rod: 39, 57, 28, 36, 25

Ask questions such as "When will the *nice rod trader* trade for another rod? How many cubes must you have? When will the *rod trader* take away your cubes? Why does the *rod trader* do that?"

rod trader maze • 2-3 players

You Need: Spinner as shown below, gameboard M11-M12, small marker for each player, base ten rods and small cubes.

How To Play: The shortest player begins. On a turn, the player spins the spinner and moves that number of spaces on the path in any direction. If the player lands on a multiple of ten, his/her turn is finished. If not, the player rounds the number landed on to the nearest ten (using rods and cubes as needed) and moves one space to the correct multiple of ten. The first person to reach HOME wins. The winner begins the next game if played.

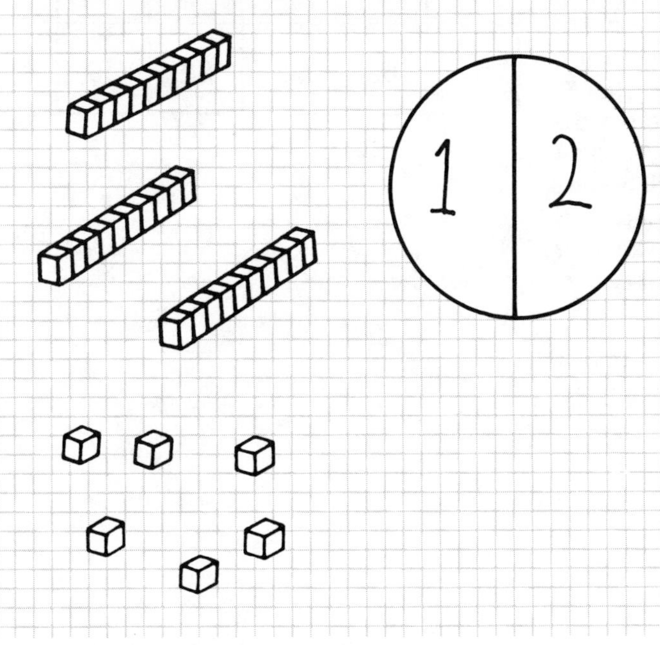

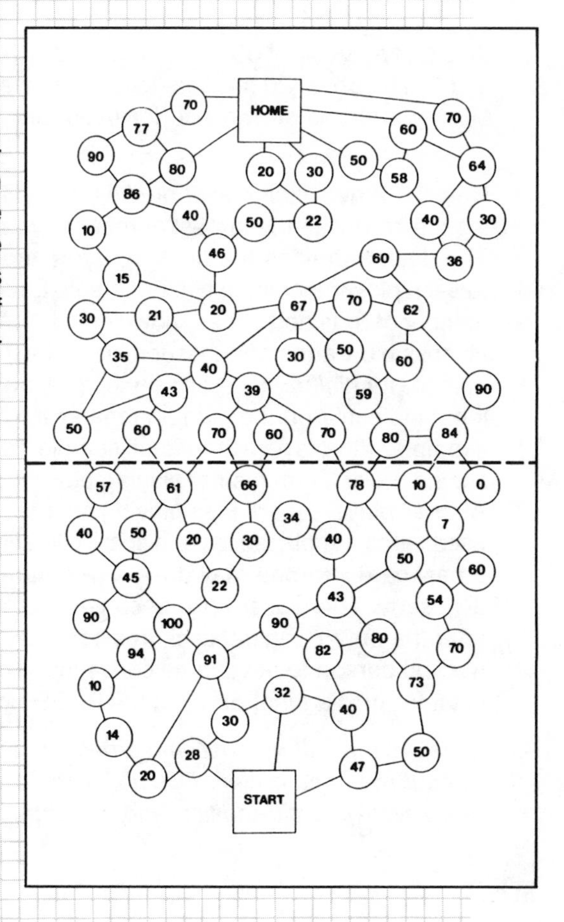

READINESS FOR TWO-DIGIT ADDITION AND SUBTRACTION

Many elementary children learn the two-digit addition and subtraction algorithm with little or no understanding and with little or no visual representation using manipulatives. If this approach is taken, the children need considerable practice with examples having no trades before moving to examples with trades. However, when teaching with materials, children will naturally add ones to ones (cubes to cubes) and tens to tens (rods with rods) and thus can move directly into solving any two-digit example. One advantage here, besides that of better understanding the algorithm, is that children do not develop a habit of not trading and are, therefore, less likely to forget to trade when needed. To make teaching this concept easier, it is helpful to practice making trades in a format similar to that used later in the algorithm.

trade • 2-5 players

T	R	A	D	E
60	42	67	43	33
44	56	31	50	55
51	40	52	54	37
57	32	45	30	61
34	63	53	62	46

You Need: Deck of 50 cards M13 and M14; base ten rods and small cubes; 20 markers for each player; playing board M8 prepared as shown one for each player.

How To Play: Each player draws a card. The player drawing the card with the largest number of cubes is the dealer. The dealer shuffles the deck of cards and deals one card to each player including him or herself. Each player gets the number of rods and cubes shown on his/her card and makes all possible trades of cubes for rods. He/she then finds the number for his/her collection on the playing board under the letter indicated on the card and marks it with a marker. For example, the player with either sample card shown below should cover 37 on his/her playing board under the letter E. After all players mark their playing boards for the round, the used cards are placed in the discard pile and all players draw a new card. No one may draw a new card until all players are ready. The discard pile becomes the playing deck only after the entire original deck has been used. The winner is the first person to have three consecutive markers in a row, column, or diagonal on his/her TRADE board.

***Note**: As the children become proficient, allow them to play without rods and cubes. This ability indicates that they are ready for two-digit addition and/or subtraction.*

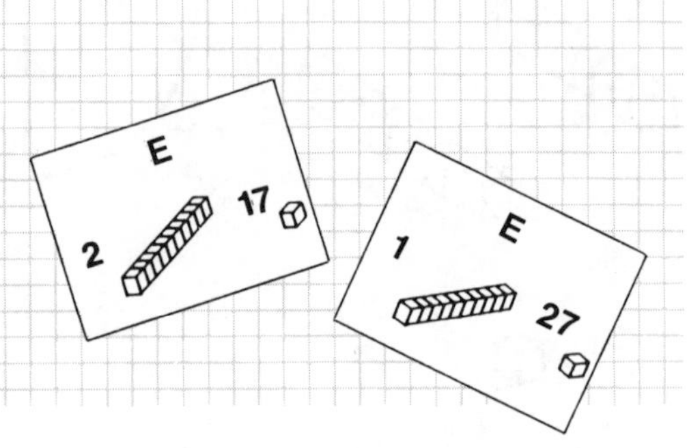

WHERE NEXT

The two-digit experiences suggested here should provide readiness for either work with the concept of one hundred and then three-digit numbers or for moving to two-digit operations. The order presented in Books 1 and 2 does not suggest one in preference to the other.

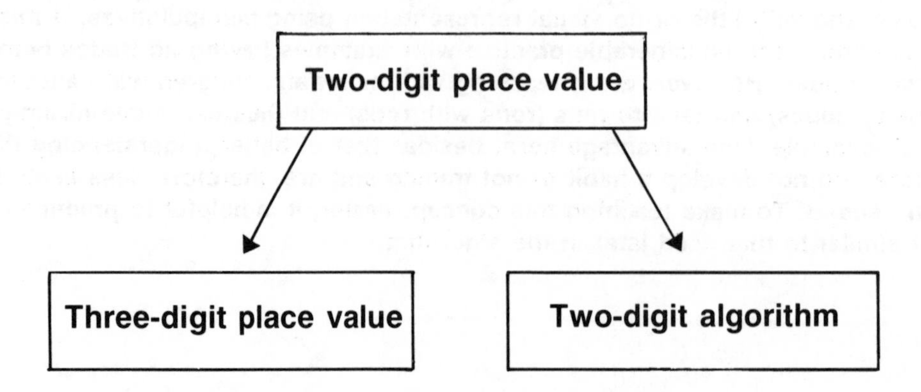

THREE-DIGIT PLACE VALUE

Many of the concepts in three-digit place value are similar to concepts in two-digit place value. It is very important for most children, however, to have extensive manipulative experiences with three-digit materials to build a foundation for understanding larger numbers. The concept of one hundred and making trades for one hundred is the basis of three-digit work.

Give each child a flat. Write the words "one hundred" and the symbol "100" on the chalkboard or overhead projector. Have the children lay rods on the flat. Ask "How many tens does it take to equal 100?" Write "10 tens" on the chalkboard or overhead projector. Have the children put aside the flat and show 90 with rods. Tell the children: "We are going to count from 90 until we can trade for 100. What comes after 90? How do we show it with rods and small cubes?" Continue to 99. Ask: "How do we show what comes next? How many cubes do we have now? What should we do? How many rods do we have after the trade? What can we do with 10 rods? What number comes after 99?" Write "next number after 99" or "1 more than 99" on the chalkboard. Have the children find other ways to make 100 such as 9 tens 10 ones, or 8 tens 20 ones.

100 tic tac toe • 2 players

You Need: "Middle" spinner M3, gameboard M15, five distinctive markers for each player.

How To Play: Players take turns spinning. The first student to spin "100" begins play. On a turn the player spins and places one of his/her markers on any unclaimed spot on the gameboard that matches the name for one hundred on the spinner. The first player to claim a row, column, or diagonal on the gameboard wins. The loser goes first on the next game played.

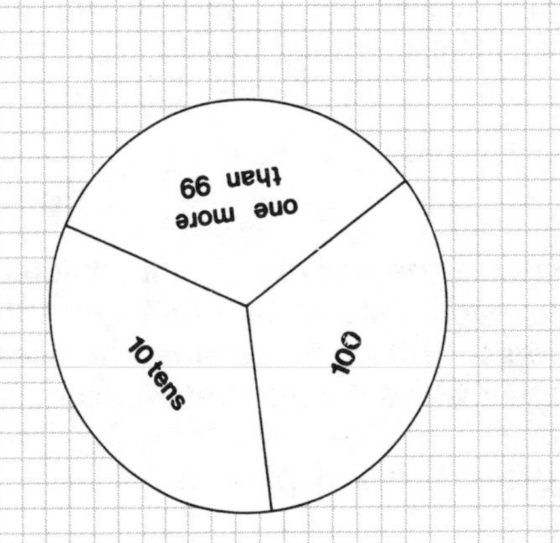

NAMING THREE-DIGIT NUMBERS

Provide students with at least 5 flats, 9 rods, and 9 small cubes. Have them show 43 with rods and cubes. Now have each student put 2 flats with the 43. Ask: "What is the number for 1 flat? Then what would we call 2 flats?" Tell the children: "2 flats, 4 rods, 3 cubes shows two hundred forty-three."

Have them show three hundred fifty-two. Walk around and check. If children have trouble, ask: "How would we show one hundred? How will we show three hundred? How will we show three hundred fifty-two?" Continue with numbers like two hundred twenty-seven and four hundred thirty-five. As the children become comfortable, write the numeral for each on the chalkboard as you say the word. Move to more difficult numbers like 101, 407, 310, 240.

As the children master showing three-digit numbers, reverse the process. Have each child lay out 2 flats, 4 rods, 7 cubes. Show a collection of your own. Ask: "What number have we shown?" Point to the flats, rods and cubes in order as you say the number together. As the children learn to say the numbers discontinue pointing. Gradually move to more difficult examples like 120, 250, and then 206, 403, 101. Continue this verbal activity until you can mix levels of examples, whether you say the number and the children show it or you tell how many flats, rods and cubes and the children show it and say the number.

little 500 • 2-5 players

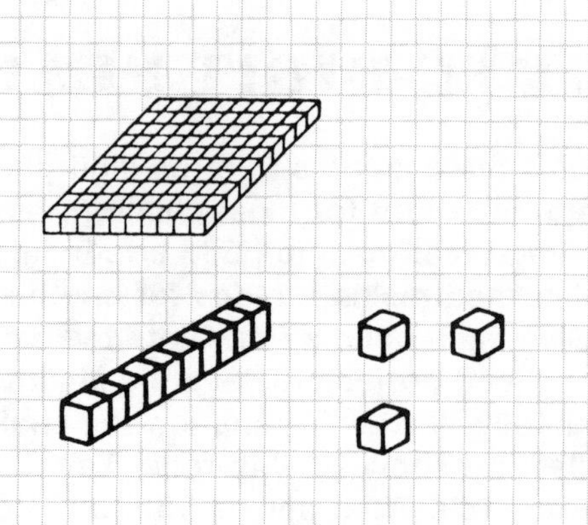

You Need: Mats M4 and M16 taped together for each player; "top" spinner M17; flats, rods and small cubes.

How To Play: Players take turns spinning. The first player to spin a flat begins. On a turn a player spins, adds to his/her collection, makes trades as possible and verbally states how many he/she has. The first player to reach 500 wins.

Extension: *Roll a die 3 times to determine how many hundreds, tens, and ones to receive on a turn. Play to 999.*

THREE-DIGIT NUMERALS

Again supply each child with flats, rods and small cubes. Have each child show 2 flats, 4 rods, 7 cubes. Have the children say the number. Write 247 on the chalkboard. Ask: "What does the 2 mean? the 4? the 7?" Repeat with other three-digit numbers. As the children become comfortable, have them lay out 3 flats, 1 rod, 6 cubes. Write 316 on the chalkboard. As you ask the questions, write "hundreds", "tens", and "ones" under the appropriate digits.

3	1	6
hundreds	**tens**	**ones**

Continue the activity. As the children seem to understand, have the children say the number and have each child write it on paper. Walk around and check. When the children have mastered this, reverse. Write a numeral on the board and have the children show it with base ten blocks. Do examples without zeros before examples with zeros.

draw II • 2-4 players

You Need: 20 picture cards selected from M18-M22 and 20 numeral cards made to match the picture cards.

How To Play: Played like DRAW (page 8).

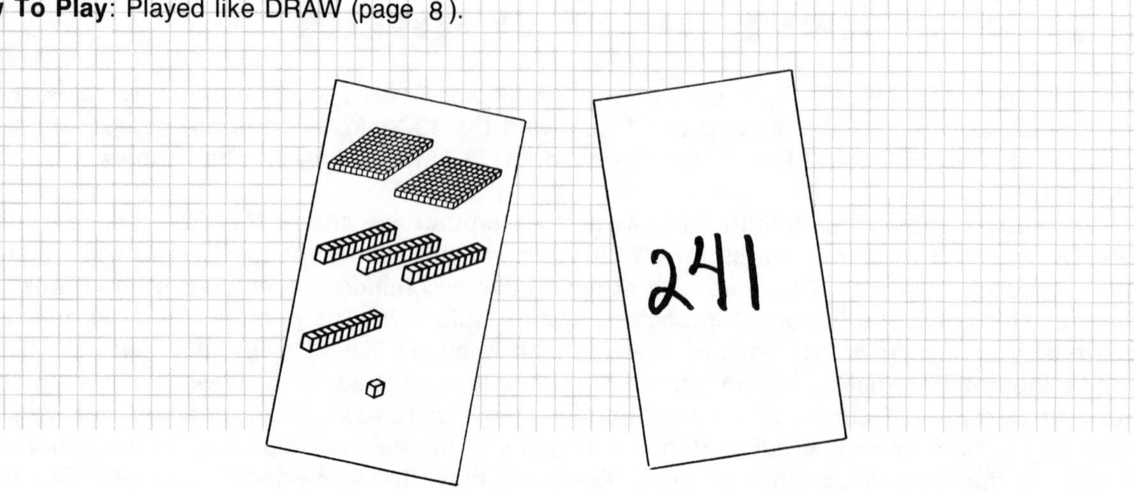

READINESS FOR THREE-DIGIT COUNTING

Finding one more or naming the number after any given three-digit number should precede counting in sequence by ones. This activity puts emphasis on the meaning of counting rather than on rote patterns.

Provide the children with flats, rods and small cubes. Write a two-digit number such as 36 on the chalkboard. Have the children show it with rods and cubes. Ask: "What is the number after 36 when counting by ones? How do we show this with the blocks?" Continue review until children remember that the next number is found by adding one cube. Write a three-digit number on the chalkboard such as 245. Have the children show it with base ten blocks. Ask: "How should we find the next number with the blocks? What is the number after 245?" Continue until the students feel comfortable, then move to more difficult examples such as the next number after 320, 250, and 470. More difficult examples include the numbers after 259, 329, and 149. Finally, the most difficult include numbers after 100, 200, 300 and so forth. Finish the lesson by mixing together examples from all levels.

chance • 2-5 players

You Need: Flats, rods and small cubes; small marker for each player; gameboard M1-M2; 50 number cards made with numbers listed below; 15-20 CHANCE cards made with messages such as "move ahead 2", "lose a turn", and "move forward to join the closest player".

Numbers: 98, 99, 100, 101, 102, 103, 108, 109, 110, 111, 112, 113, 124, 125, 126, 127, 128, 129, 130, 198, 199, 200, 201, 202, 224, 225, 226, 236, 247, 248, 300, 301, 306, 309, 310, 314, 355, 356, 357, 358, 359, 360, 399, 400, 401, 409, 410, 411, 412, 413.

How To Play: Players draw a random card in turn. The first player to draw a number in the two hundreds begins play. Shuffle number cards and place face down. Shuffle CHANCE cards and place face down. On a turn a player draws a number card, shows the number with base ten blocks, finds the next number (trading, if necessary) and says the two numbers aloud. The player then moves ahead on the gameboard the number of ones in his/her collection. If he/she has zero ones, the player draws a CHANCE card and follows the message. The first person to reach HOME wins the game.

after • 2-4 players

You Need: M8 playing boards, prepared as shown, one for each player; deck of 50 picture cards M18-M22 labeled as indicated; 10 to 20 markers for each player.

To prepare playing boards, label as shown.

(1)

A	F	T	E	R
101	111	257	260	404
102	242	300	261	406
103	245	301	262	409
104	246	302	328	412
110	248	305	331	413

(2)

A	F	T	E	R
101	111	302	260	404
104	112	303	328	405
105	113	304	329	407
106	243	306	330	410
108	248	358	332	412

(3)

A	F	T	E	R
102	112	301	259	406
105	242	303	261	407
107	243	305	329	408
108	244	306	332	411
109	247	307	333	413

(4)

A	F	T	E	R
103	113	257	259	405
106	244	258	262	408
107	245	300	330	409
109	246	304	331	410
110	247	307	333	411

To prepare cards, write a letter and a number on each one.

A: 100, 101, 102, 103, 104, 105, 106, 107, 108, 109
F: 110, 111, 112, 241, 242, 243, 244, 245, 246, 247
T: 299, 300, 301, 302, 303, 304, 305, 306, 256, 257
E: 258, 259, 260, 261, 327, 328, 329, 330, 331, 332
R: 403, 404, 405, 406, 407, 408, 409, 410, 411, 412

How To Play: Shuffle picture cards and lay face down. On a turn, a player turns the top card up for all to see and says the number. Each player figures out the next whole number and covers it on his/her playing board if it is there under the letter on the card. For example, if the picture card for R 404 is turned up, R 405 may be covered on boards 2 and 4. The first person to cover a complete row, column, or diagonal wins.

***Extension**: Use numeral cards instead of picture cards.*

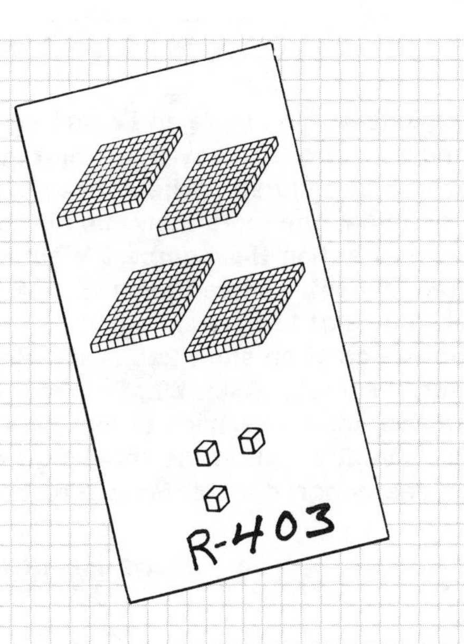

COUNTING BY ONES

Having practiced one more than any number, the children should be ready to count orally, then write numerals in order starting at any three-digit number. (Example: Count from 200 to 300.) If the children have difficulty, have them use the base ten blocks as they say the numbers in sequence. The numbers such as those from 100 to 112 or from 200 to 212 seem to be the most difficult. It is sometimes helpful to use the blocks to get started after each multiple of one hundred.

skip and count • 2-4 players

You Need: Deck of 50 picture cards M18-M22 with numbers shown written on each; 15 cards with the word "SKIP" written on each.

How To Play: Shuffle picture cards and SKIP cards together. Deal 6 cards to each player. The player with the lowest numeral card begins play. On a turn a player may play one or more cards. To play a card the person must play the next number when counting by ones or begin a new sequence by playing SKIP card followed by one or more picture cards. After each turn the player draws the top card from the deck. If he/she can't play either the next number or a SKIP card, the player loses his/her turn and must draw the top card from the deck. The first person to use all the cards in his/her hand or the person holding the fewest cards when the deck is gone and no further cards can be played, wins.

***Extension**: Play with numeral cards instead of picture cards.*

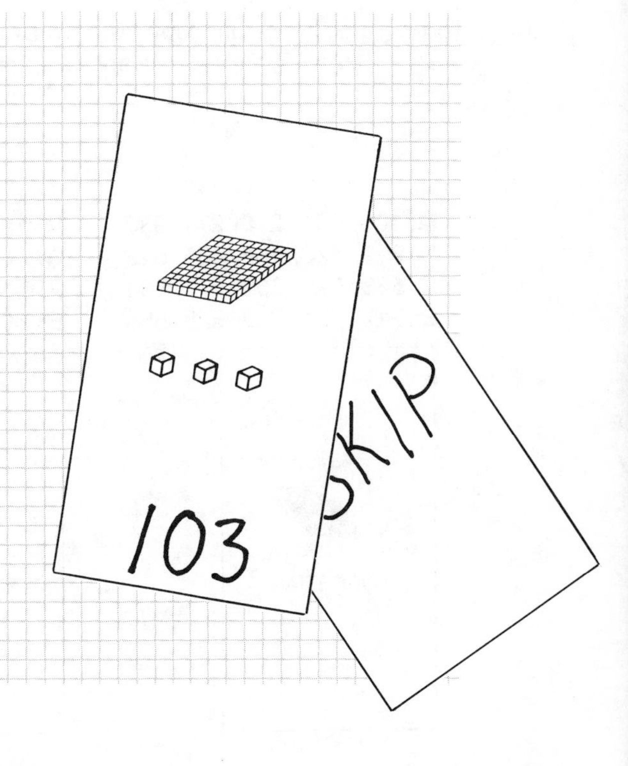

TEN MORE AND ONE HUNDRED MORE

Having mastered one more and counting by ones, the students are ready to learn ten more or one hundred more than any three-digit number. Provide flats, rods and small cubes for each child or pair of children. Write a three-digit number such as 321 on the chalkboard. Ask: "How would we show one more than this number? What is one more than 321? Then how shall we show ten more than this number? What is ten more than 321?" Have the children show 135. Now have them show ten more than 135. Ask: "What number is ten more than 135?" Do three or four more examples of ten more.

Have the children show 215. Ask: "How shall we show one hundred more than 215?" Have the children show it. Ask: "What number is one hundred more than 215?"

Continue doing examples of ten more and one hundred more. As the children become comfortable, have them show the three-digit number and visualize (rather than actually show) ten more or one hundred more. Make a record and look for patterns.

One Hundred More		Ten More	
142	242	217	227
329	429	342	352
250	350	129	139

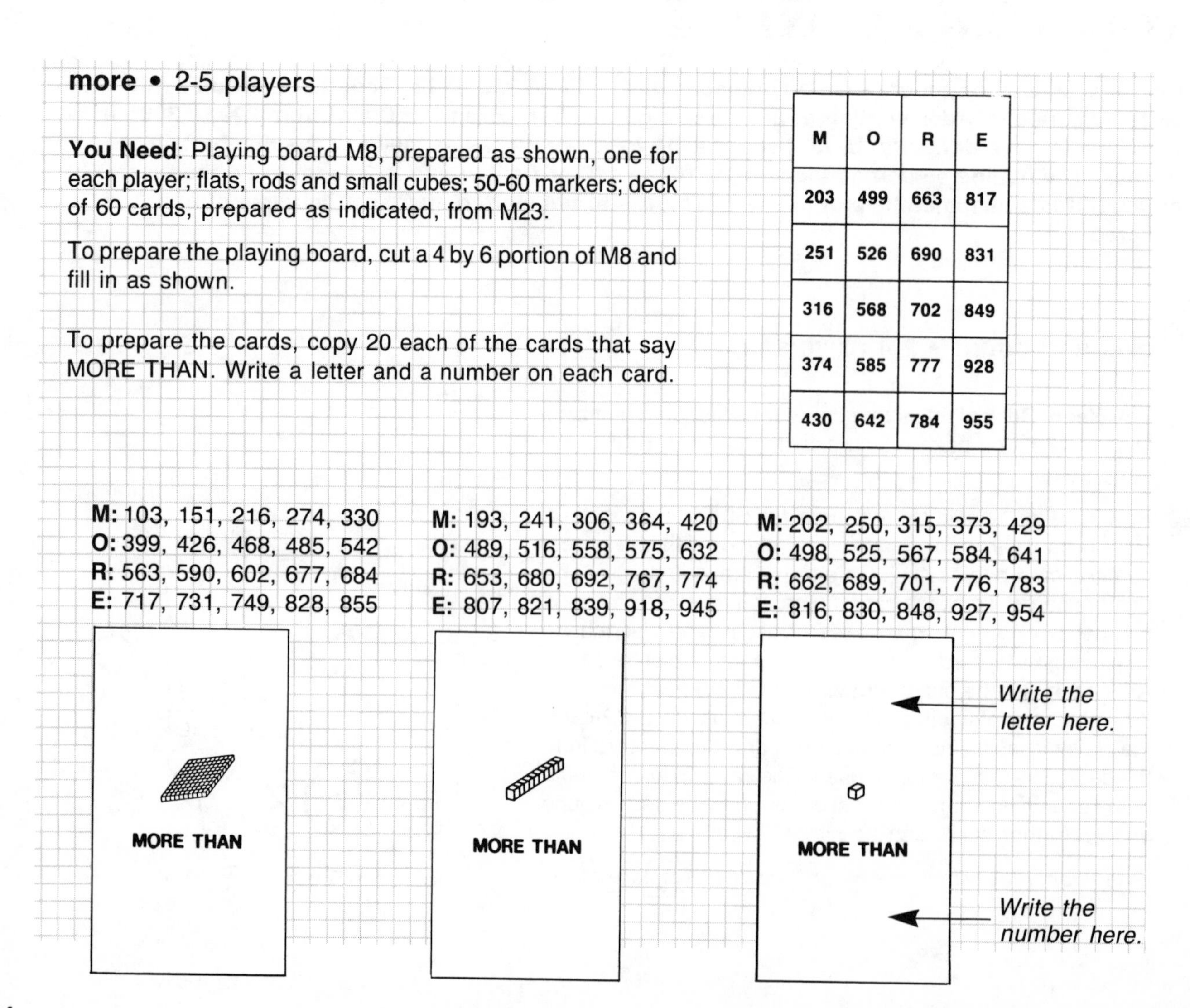

more • 2-5 players

You Need: Playing board M8, prepared as shown, one for each player; flats, rods and small cubes; 50-60 markers; deck of 60 cards, prepared as indicated, from M23.

To prepare the playing board, cut a 4 by 6 portion of M8 and fill in as shown.

To prepare the cards, copy 20 each of the cards that say MORE THAN. Write a letter and a number on each card.

M	O	R	E
203	499	663	817
251	526	690	831
316	568	702	849
374	585	777	928
430	642	784	955

M: 103, 151, 216, 274, 330
O: 399, 426, 468, 485, 542
R: 563, 590, 602, 677, 684
E: 717, 731, 749, 828, 855

M: 193, 241, 306, 364, 420
O: 489, 516, 558, 575, 632
R: 653, 680, 692, 767, 774
E: 807, 821, 839, 918, 945

M: 202, 250, 315, 373, 429
O: 498, 525, 567, 584, 641
R: 662, 689, 701, 776, 783
E: 816, 830, 848, 927, 954

How To Play: Each player draws a card, shows the number with flats, rods and cubes and puts a flat, rod or cube with the collection as indicated on the card. Each player finds the number for this collection on his/her playing board under the letter on the card and marks it with a marker. All used cards are then discarded and all players draw a new card. No one may draw a new card until all players are ready.If a player draws a card for a number already covered on his/her gameboard, that player does not cover it a second time and must wait until everyone is ready before drawing another card. Play continues until one person has three consecutive markers in a row, column, or diagonal.

***Extension**: Prepare cards with only letters and numerals as shown. The child must declare if the number played on the gameboard is one hundred more, ten more or one more.*

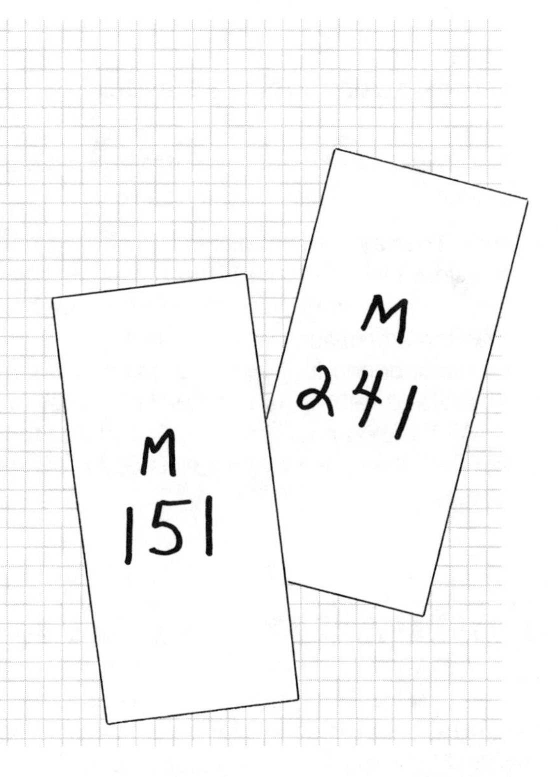

***Note**: Allow children to play without base ten blocks as they are ready.*

more rummy • 2-5 players

You Need: Picture cards from MORE and a card for each number on the MORE playing board.

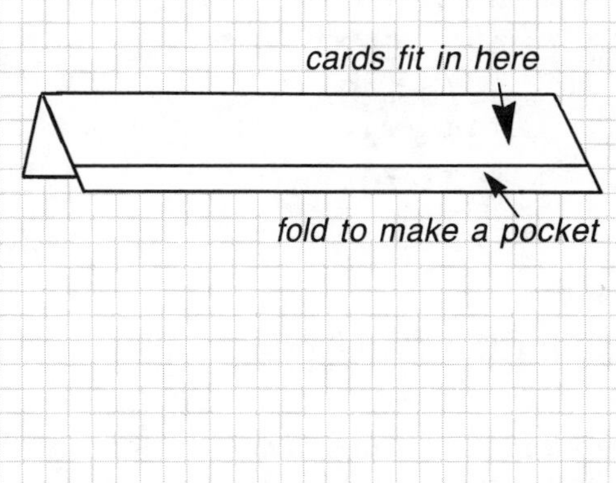

How To Play: The dealer shuffles all cards and deals 5 to each player. Turn over the top card to begin a discard deck. On a turn a player may draw one card from the face down deck or up to four cards from the discard deck. The player then tries to make a book of 3 or 4 related cards and/or play the fourth card on another player's book of three. After each turn the player must discard one card. The winner is the first player to get rid of all the cards in his/her hand.

A sample book:

R	R	R	R
MORE THAN 677	MORE THAN 767	MORE THAN 776	777

more-fewer race • 2 players

You Need: "Bottom" spinner M17, playing board M24, two distinctive markers for each player.

How To Play: One marker is placed on each START HERE position. On a turn the player spins and moves one space forward, diagonally, or sideward to a number more or fewer than the amount shown on the spinner. If a space is occupied by the opponent, a player may place his/her marker on that space and return the opponent's marker to one of his/her START HERE positions. The first player to have both markers anywhere in the opponent's START HERE row wins the game.

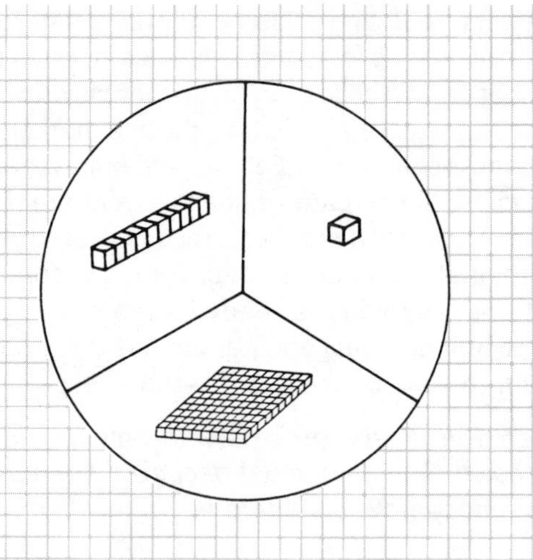

COMPARING THREE-DIGIT NUMBERS

Let the children work in pairs. Write 39 and 43 on the board. Have one child in each pair show 39 and the other child show 43. Ask: "Which number is larger? How can we tell using the blocks?" Do more two-digit examples for review as needed. Write 423 and 502 on the board. Have the children show with base ten blocks. Ask: "Which is larger? How can you tell?" Continue comparing using base ten blocks until the children seem to master the concept.

Record:

$423 < 502$
$357 > 149$
$729 < 735$
$143 > 76$

rod war II • 2 players

You Need: 20 picture cards selected from M18-M22 with number shown written on each, reproduced three times, for a total of 60 cards; flats, rods, and small cubes.

How To Play: Played like ROD WAR (page 13).

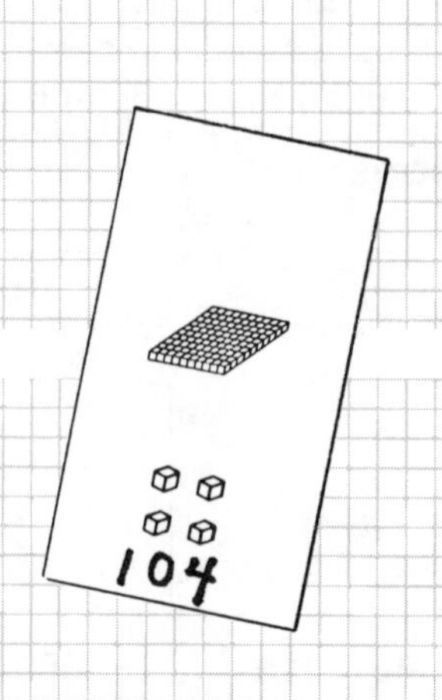

four in a row II • 2 players

You Need: 20 picture cards selected from M18-M22 with the number shown written on each.

How To Play: Played like FOUR IN A ROW (page 13).

ROUNDING TO TENS AND HUNDREDS

Provide the children with flats, rods and small cubes. Have the children show 48. Ask them if they remember the *nice rod trader* who would sometimes trade for a rod even when there were not quite enough cubes. If they remember, ask: "What else did the *rod trader* do? What would the *rod trader* do with 48?" Continue reviewing rounding two-digit numbers to tens. Have the children show 286. Say: "We want to round to hundreds. What base ten blocks do we want? Will the *rod trader* give us another flat? Why do you think so?" Have the children show 113. Ask: "What pieces of wood do we want if we round to hundreds? Will the *rod trader* give us another flat? Why?" Continue with other quite obvious examples such as 190, 379, 206, and 120. Remind the children that the *nice rod trader* gave them another rod if they had at least 5 cubes or half a rod. Ask: "How much is half a flat?" Let them figure it out with base ten blocks. Ask: "What will the *nice rod trader* do with 260? Why do you think that?" Continue showing three-digit numbers and rounding to hundreds.

100's round-up • 2-4 players

You Need: Playing board, as shown, one for each player; a deck of 32 numeral cards marked with one of the numbers listed below; flats, rods and small cubes; 20 to 30 markers.

Playing board:

100	200	300	400	500	600	700	800	900

Numbers: 103, 120, 145, 160, 187, 215, 236, 254, 298, 310, 327, 363, 390, 409, 431, 450, 475, 512, 540, 583, 637, 648, 721, 730, 760, 777, 809, 830, 872, 895, 925, 942.

How To Play: Shuffle numeral cards and lay face down. The tallest player begins play. On a turn a player draws a numeral card, shows the number with rods, flats and cubes, rounds it to the nearest hundred and covers that number on his/her gameboard with a marker. If a player needs a number that has already been covered, he/she loses a turn. The first person to get four consecutive spaces covered wins.

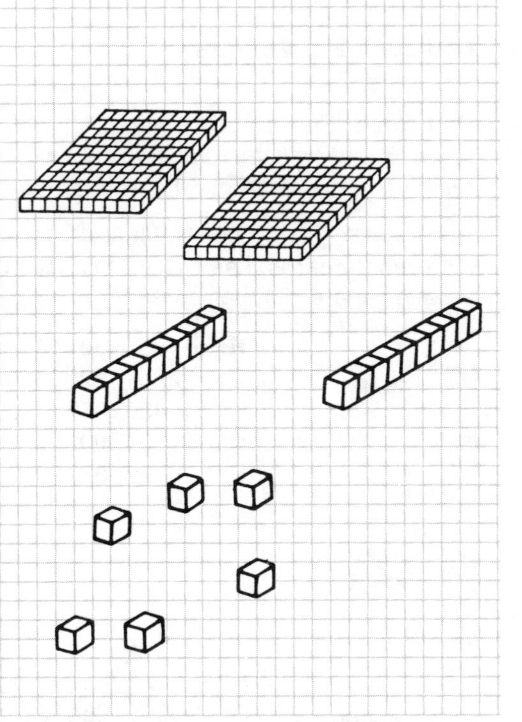

Review rounding two-digit numbers to tens and three-digit numbers to hundreds. Write 279 on the board. Have the students show it with flats, rods and cubes. Ask: "When rounding to tens what base ten blocks do we want? When rounding to hundreds, what base ten blocks do we want? What is 279 rounded to the nearest hundreds? Can you figure out what 279 would be rounded to the nearest tens? Could the 2 flats be traded for rods? Would we need a *nice rod trader* to do this? Would I have the same number with 20 rods or 2 flats?" Say: "Since I have the same number, let's leave them as flats and save some work." Ask: "What should we do with 7 rods? What should we do with 9 cubes? What number do we have?"

Write on the board:

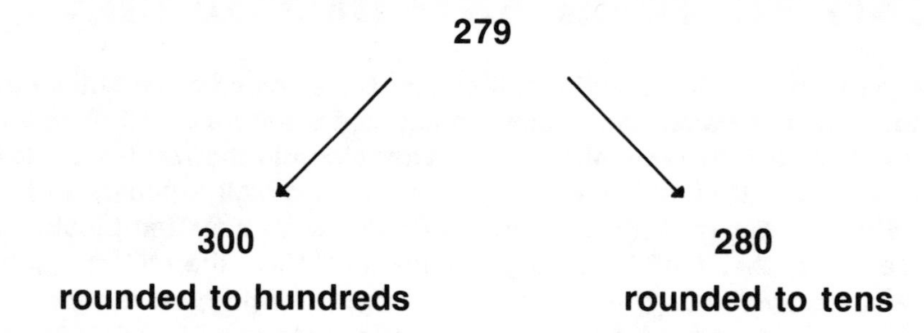

Continue with other examples asking questions as needed until the children can round 3-digit numbers to hundreds and to tens. The most difficult examples are rounding numbers such as 296 and 304 to the nearest tens and nearest hundreds.

round * 2-5 players

You Need: Playing board M8, prepared as indicated, one for each player; flats, rods, and cubes; 20 to 30 markers for each player; a deck of 50 cards, prepared as indicated.

To prepare playing board, fill in as shown.
To prepare cards, use ROUND cards from M23 and write a letter and a number on each as indicated below.

R	O	U	N	D
100	260	450	600	780
130	300	490	620	800
170	310	500	660	870
200	340	550	700	900
220	400	580	730	940

ROUND

TO

R: 85, 130, 186, 215
O: 267, 321, 359, 423
U: 490, 536
N: 594, 602, 677, 737
D: 769, 872, 907

ROUND

TO

R: 126, 132, 167, 174, 219, 222
O: 255, 264, 308, 312, 337, 344
U: 446, 451, 488, 493, 549, 550, 575, 582
N: 619, 624, 655, 661, 727, 733
D: 779, 784, 801, 868, 872, 939, 941

How To Play: Played like TRADE (page 18).

After the children have played ROUND, do more examples with the base ten blocks, looking for patterns and trying to lead the children to the "rules".

READINESS FOR THREE-DIGIT ADDITION AND SUBTRACTION

Like two-digit addition and subtraction, the biggest problem with any three-digit algorithm is understanding "borrow" and/or "carry". Both of these concepts are embodied in the action of trading. Thus, the more comfortable children feel about trading, the easier the algorithm becomes.

Provide flats, rods and small cubes. Have the children show a three-digit number like 254. Ask: "How else could we show this number? What trades could we make? Does anyone have another way? Do we still have 254?" Continue with other numbers.

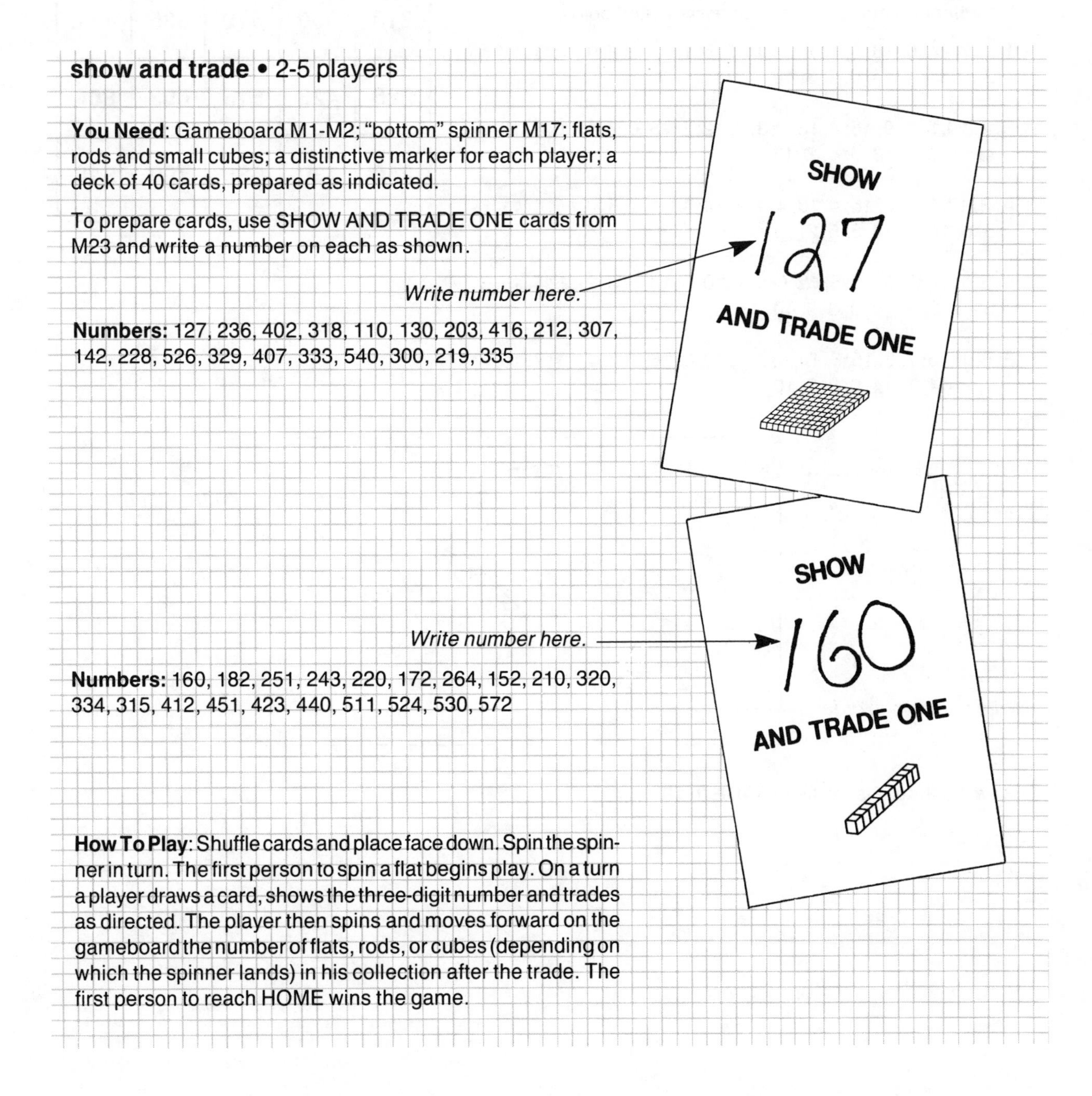

show and trade • 2-5 players

You Need: Gameboard M1-M2; "bottom" spinner M17; flats, rods and small cubes; a distinctive marker for each player; a deck of 40 cards, prepared as indicated.

To prepare cards, use SHOW AND TRADE ONE cards from M23 and write a number on each as shown.

Write number here.

Numbers: 127, 236, 402, 318, 110, 130, 203, 416, 212, 307, 142, 228, 526, 329, 407, 333, 540, 300, 219, 335

Write number here.

Numbers: 160, 182, 251, 243, 220, 172, 264, 152, 210, 320, 334, 315, 412, 451, 423, 440, 511, 524, 530, 572

How To Play: Shuffle cards and place face down. Spin the spinner in turn. The first person to spin a flat begins play. On a turn a player draws a card, shows the three-digit number and trades as directed. The player then spins and moves forward on the gameboard the number of flats, rods, or cubes (depending on which the spinner lands) in his collection after the trade. The first person to reach HOME wins the game.

trade II • 2-5 players

T_{II}	R_{II}	A_{II}	D_{II}	E_{II}
184	305	475	602	715
196	332	504	620	740
213	350	510	635	751
225	423	515	653	804
240	430	535	700	840

You Need: Playing board M8, prepared as shown, one for each player; about 20 markers for each player; flats, rods and small cubes; a deck of 50 cards, prepared as shown.

To prepare playing board, label as indicated.

To prepare cards, use the form without a title from M23 and write the letter and numbers on each as indicated below.

T: 1, 7, 14; 0, 18, 4; 1, 8, 16; 0, 19, 6; 1, 11, 3; 2, 0, 13; 1, 12, 5; 2, 1, 15; 1, 14, 0; 2, 3, 10

R: 2, 10, 5; 2, 9, 15; 2, 13, 2; 3, 2, 12; 2, 15, 0; 3, 4, 10; 3, 12, 3; 4, 1, 13; 3, 13, 0; 4, 2, 10

A: 3, 17, 5; 4, 6, 15; 4, 10, 4; 4, 9, 14; 4, 11, 0; 5, 0, 10; 4, 11, 5; 5, 0, 15; 4, 13, 5; 5, 2, 15

D: 5, 10, 2; 5, 9, 12; 5, 12, 0; 6, 1, 10; 5, 13, 5; 6, 2, 15; 5, 15, 3; 6, 4, 13; 6, 10, 0; 6, 9, 10

E: 6, 11, 5; 7, 0, 15; 6, 14, 0; 7, 3, 10; 6, 15, 1; 7, 4, 11; 7, 10, 4; 7, 9, 14; 7, 14, 0; 8, 3, 10

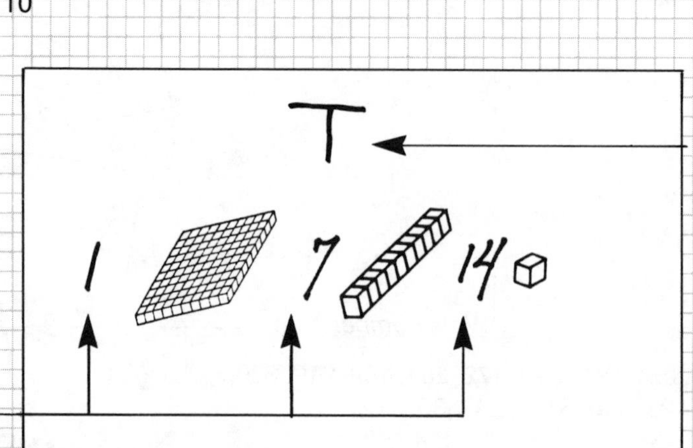

How To Play: Played like TRADE (p.18)

ON TO FOUR-DIGIT NUMBERS

Many of the games and activities for two and three-digit numbers may be expanded and modified for four-digit numbers. However, special care must be taken in introducing the concept of one thousand when beginning to explore formally the pattern of ten to one trades that govern all work in place value.

Provide each pair or small group of children with at least 1 large cube, 10 flats, 20 rods and 20 small cubes. Have the children hold up a flat. Ask: "What name do we call this?" If they answer "flat" accept as correct but continue questioning until you arrive at "hundred" or "one hundred". Do the same for a rod and a small cube.

Record.

___	___	___
hundreds	tens	ones

Have the students hold up a large cube. Tell them that this is our model of one thousand. Ask: "Where should I put the thousands place here on the board? Why?" Have the children build a stack of hundreds to match the thousand cube. Ask: "How many hundreds does it take to make one thousand?"

Record.

one thousand (1000) = 10 hundreds

Ask: "How many tens does it take to make one hundred? So, how many tens will it take to make one thousand? How could we figure this out?" After the children have figured it out, record.

one thousand (1000) = 10 hundreds
one thousand (1000) = 100 tens

Ask the children: "What is the next number after 347? Show it with base ten blocks." Ask: "What is the next number after 600? 758? 879? 900? 910? 950?969? 990? 995? 998? 999?" Have them show each with base ten blocks, trading as necessary. When they have found the next number after 999, continue to record:

one thousand (1000) = 10 hundreds
one thousand (1000) = 100 tens
one thousand (1000) = next counting number after 999

Ask the children: "Are there other ways we could show one thousand?" Record as they explore.

trading up to 1000 * 2-4 players

You Need: Playing mats M4 and M16 taped together for each player; large cube, flats, rods and small cubes; 3 dice (preferably different colors).

How To Play: Have each child roll one die. The person rolling the largest number begins play. On a turn a player rolls three dice to determine the number of ones, tens and hundreds he is to receive. On subsequent turns he adds to his collection (as in TRADING UP) making trades when possible. The first player to trade for the thousand cube wins the game.

After playing the game several times, again put the place value words on the board:

______	______	___	___
thousands	**hundreds**	**tens**	**ones**

Ask: "How many ones trade for a ten? How many trade for a hundred? How many hundreds trade for one thousand? How many thousands do you think will trade for the next place? Why do you think we always trade ten for one? (See your fingers.) Do you think everyone has always traded ten for one?" This might be an interesting place for a brief look at historic numeration systems.

APPENDIX

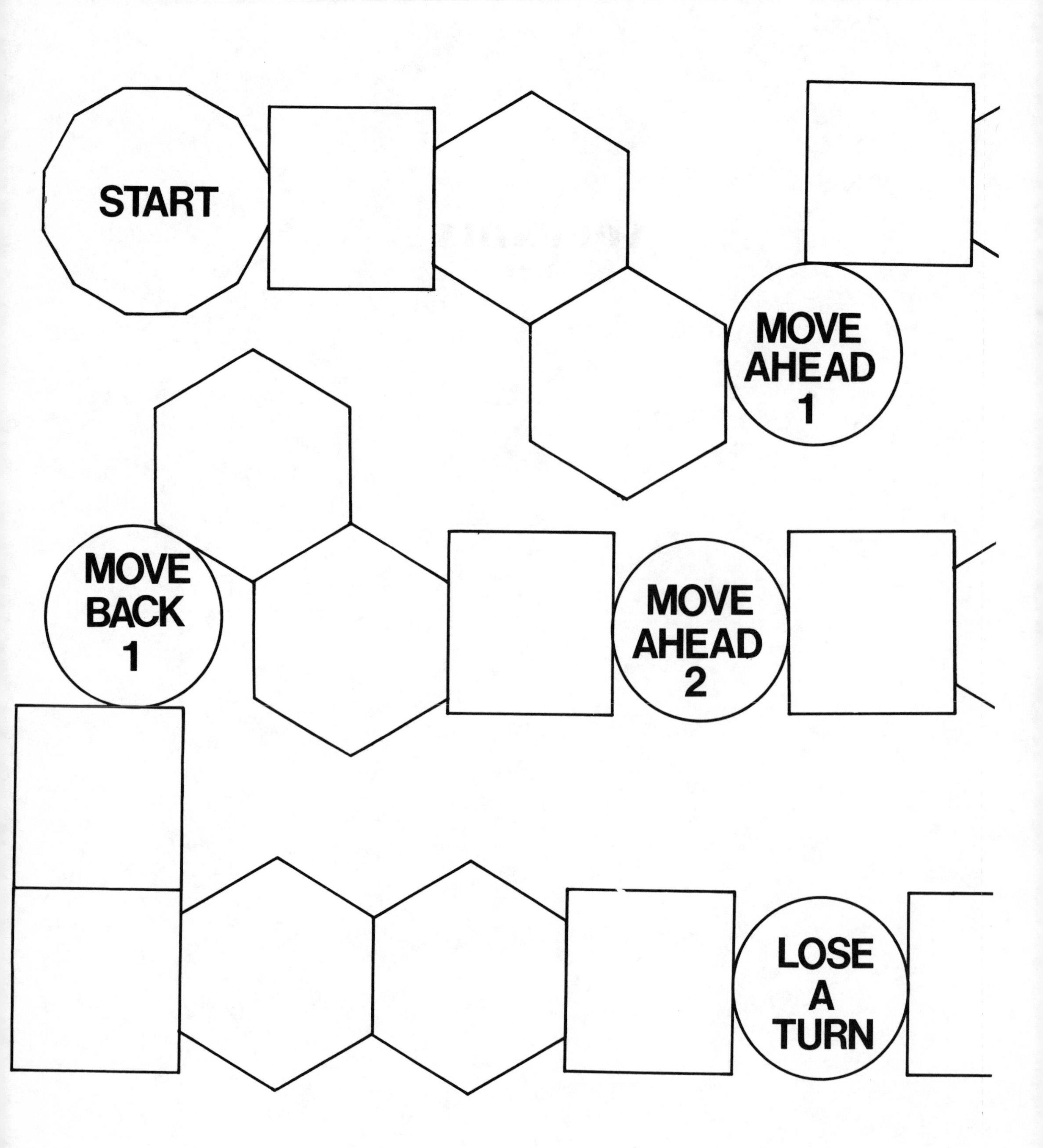
START
MOVE AHEAD 1
MOVE BACK 1
MOVE AHEAD 2
LOSE A TURN

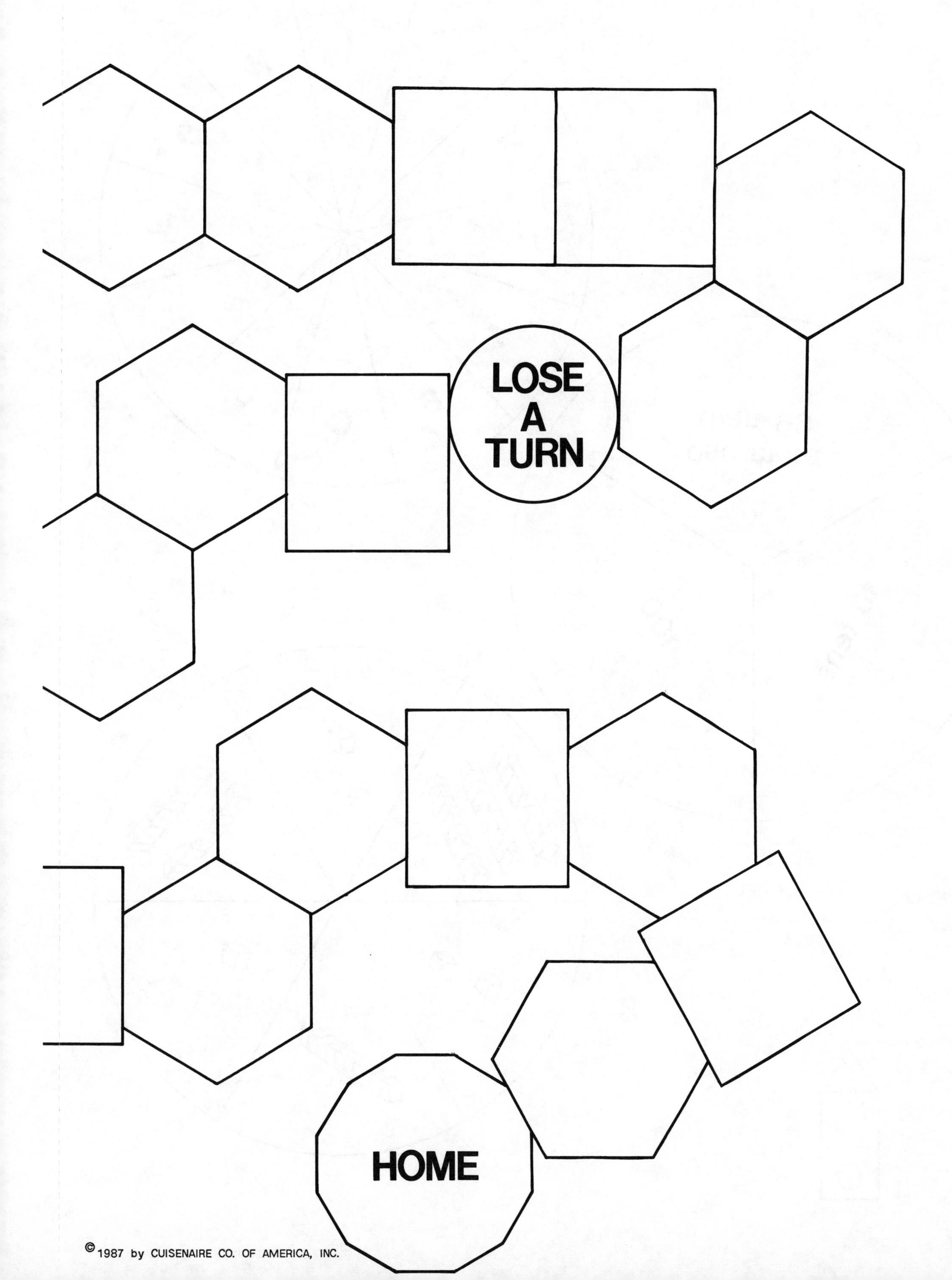
LOSE
A
TURN
HOME

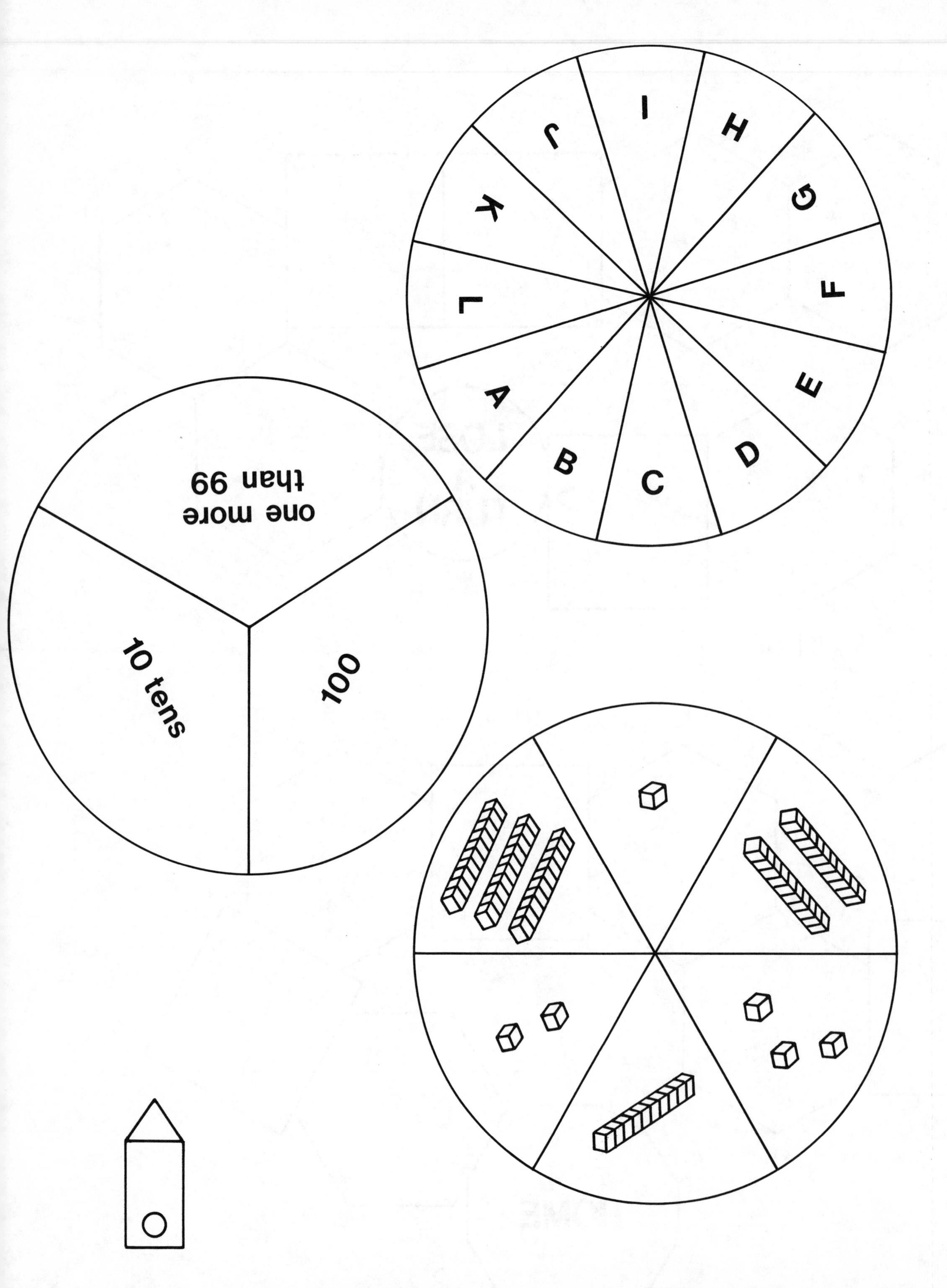
I
H
G
F
E
D
C
B
A
L
K
J
one more than 99
10 tens
100

tens

ones

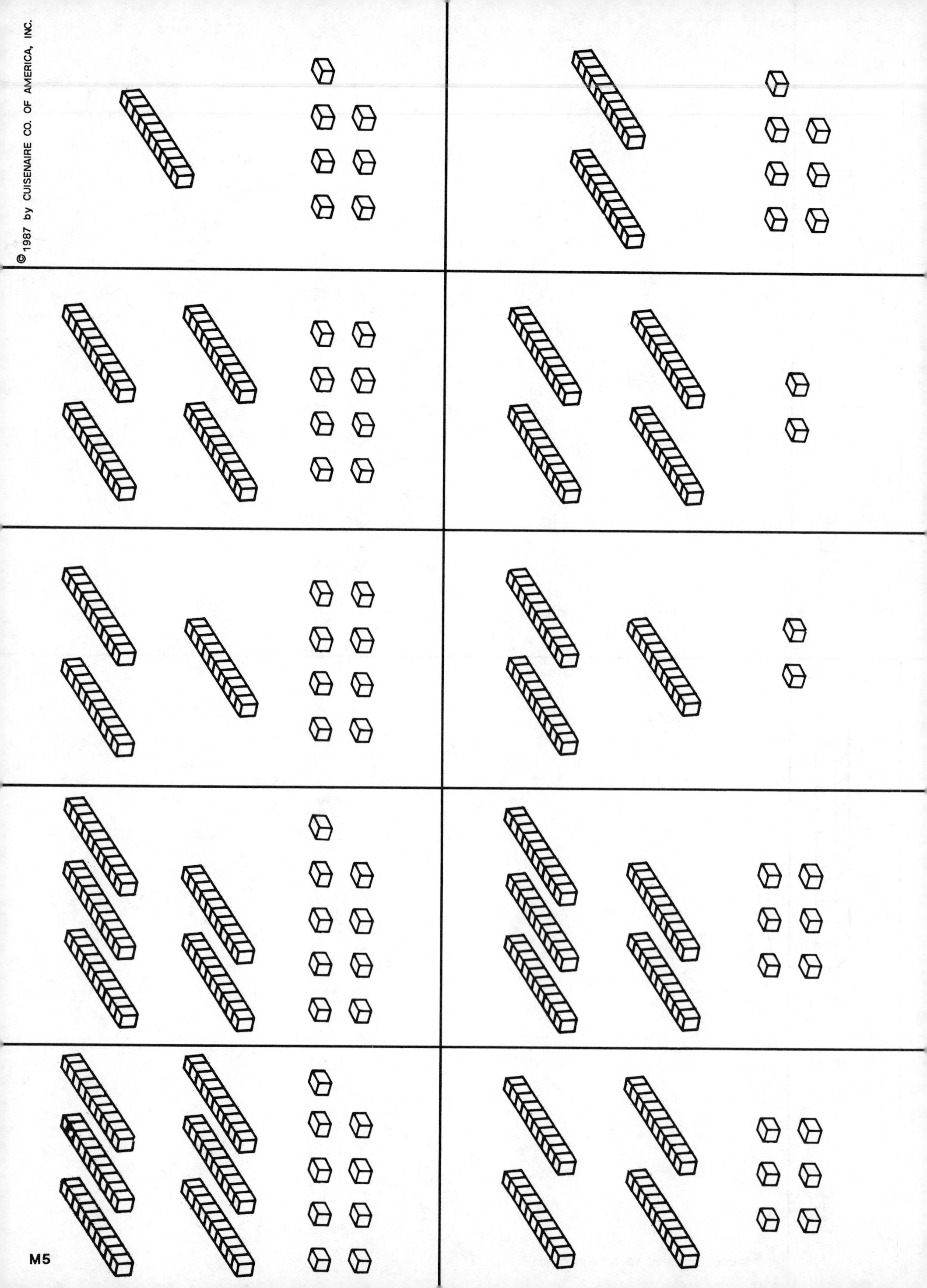
M5

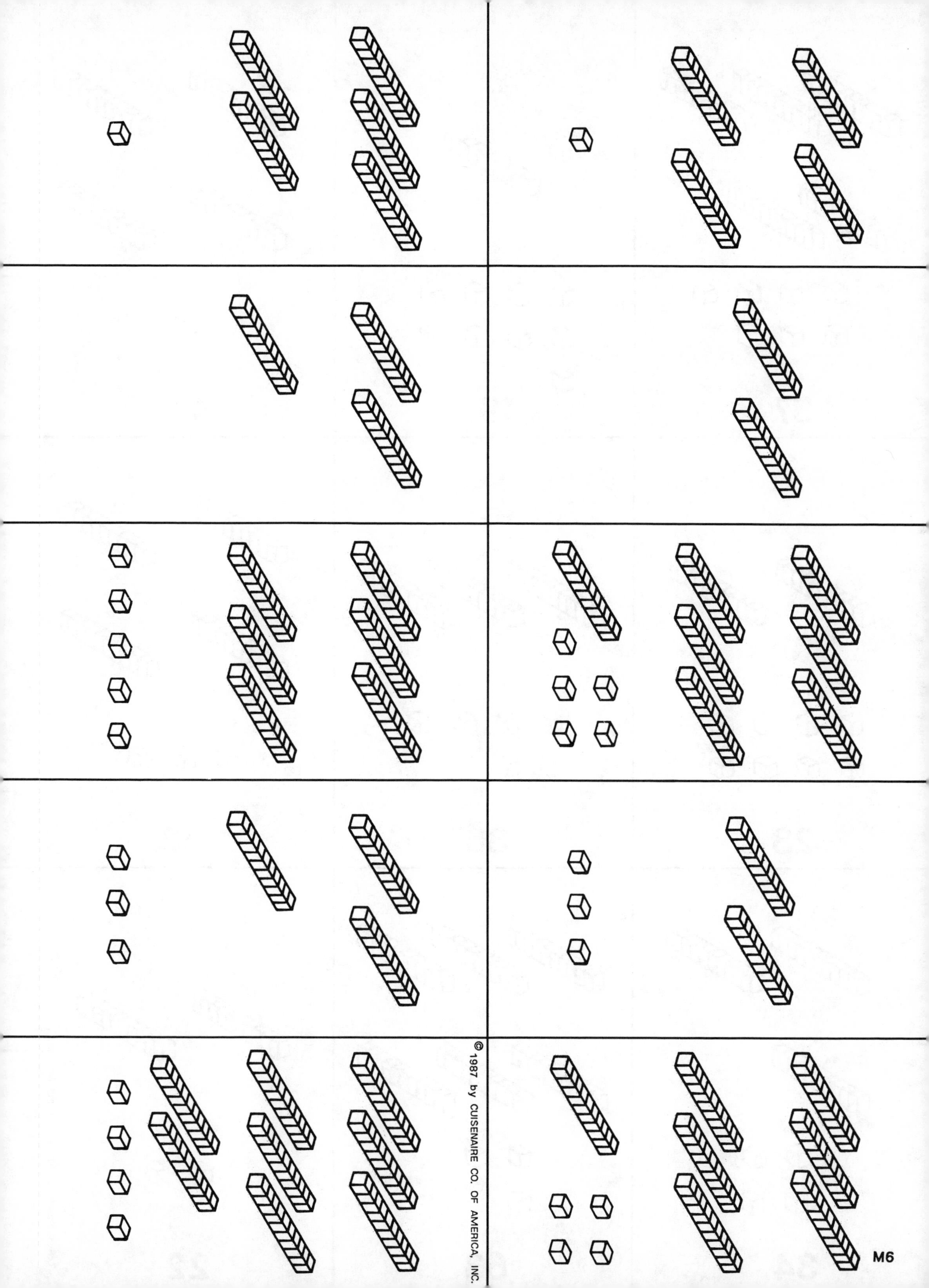
M6

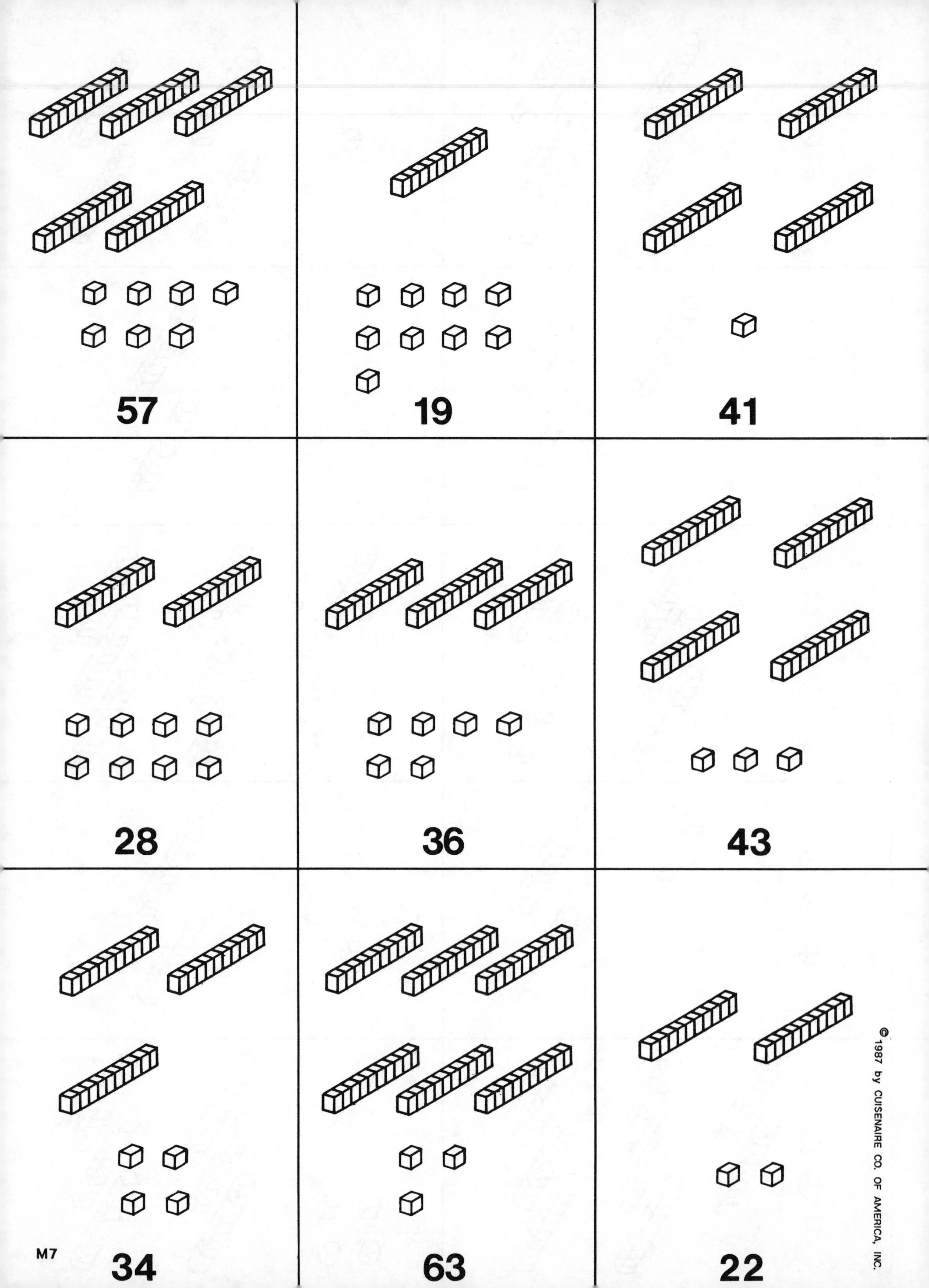
57
19
41
28
36
43
34
63
22
M7

START

1	2	3	4	5
11	12	13	14	15
21	22	23	24	25
31	32	33	34	35
41	42	43	44	45
51	52	53	54	55
61	62	63	64	65
71	72	73	74	75
81	82	83	84	85
91	92	93	94	95

6	7	8	9	10
16	17	18	19	20
26	27	28	29	30
36	37	38	39	40
46	47	48	49	50
56	57	58	59	60
66	67	68	69	70
76	77	78	79	80
86	87	88	89	90
96	97	98	99	END

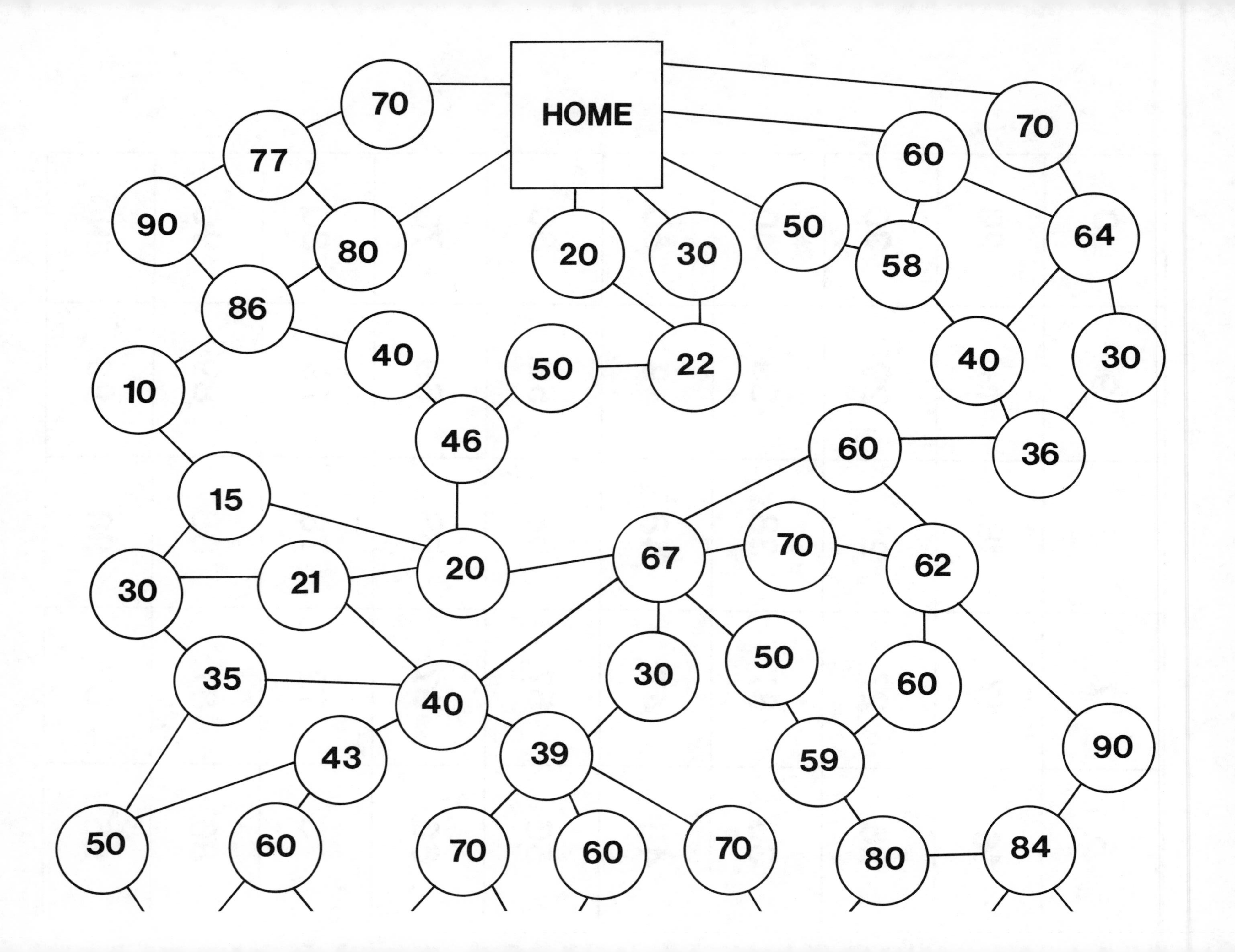
HOME
70
77
90
80
86
10
40
46
50
20
30
22
50
60
58
70
64
40
30
36
60
15
30
21
20
67
70
62
35
40
30
50
60
43
39
59
90
50
60
70
60
70
80
84

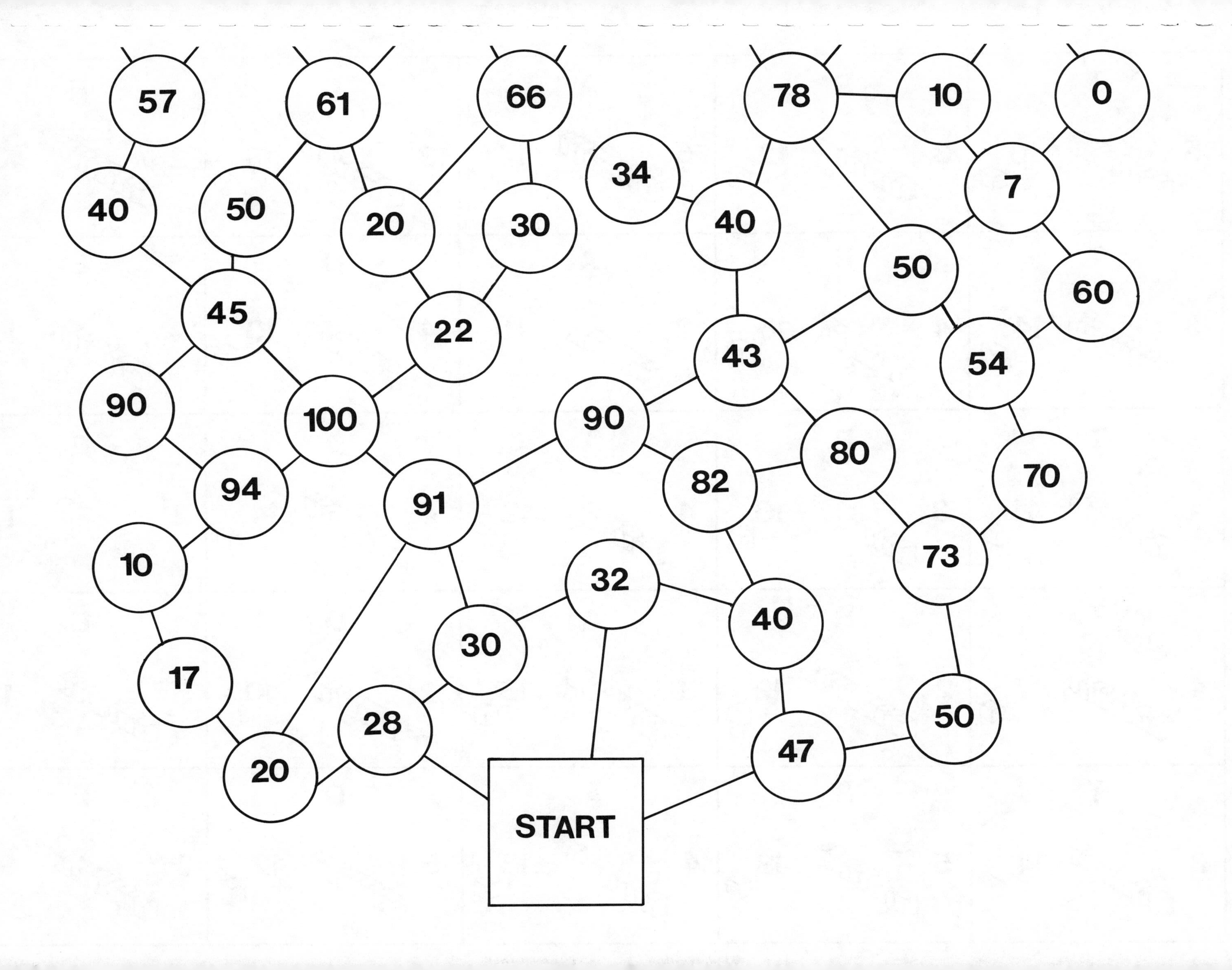
57
61
66
78
10
0
40
50
20
30
34
40
7
50
60
45
22
43
54
90
100
90
80
70
94
91
82
73
10
32
40
30
17
50
28
47
20
START

T	R	A	D	E
5 10	3 12	5 17	3 13	2 13
3 14	4 16	2 11	4 10	4 15
4 11	3 10	4 12	4 14	2 17
4 17	2 12	3 15	2 10	5 11
2 14	5 13	4 13	5 12	3 16

M14

T	R	A	D	E
4 20	2 22	4 27	2 23	1 23
2 24	3 26	1 21	3 20	3 25
3 21	2 20	3 22	3 24	1 27
3 27	1 24	2 25	1 20	4 21
1 24	4 23	3 23	4 22	2 26

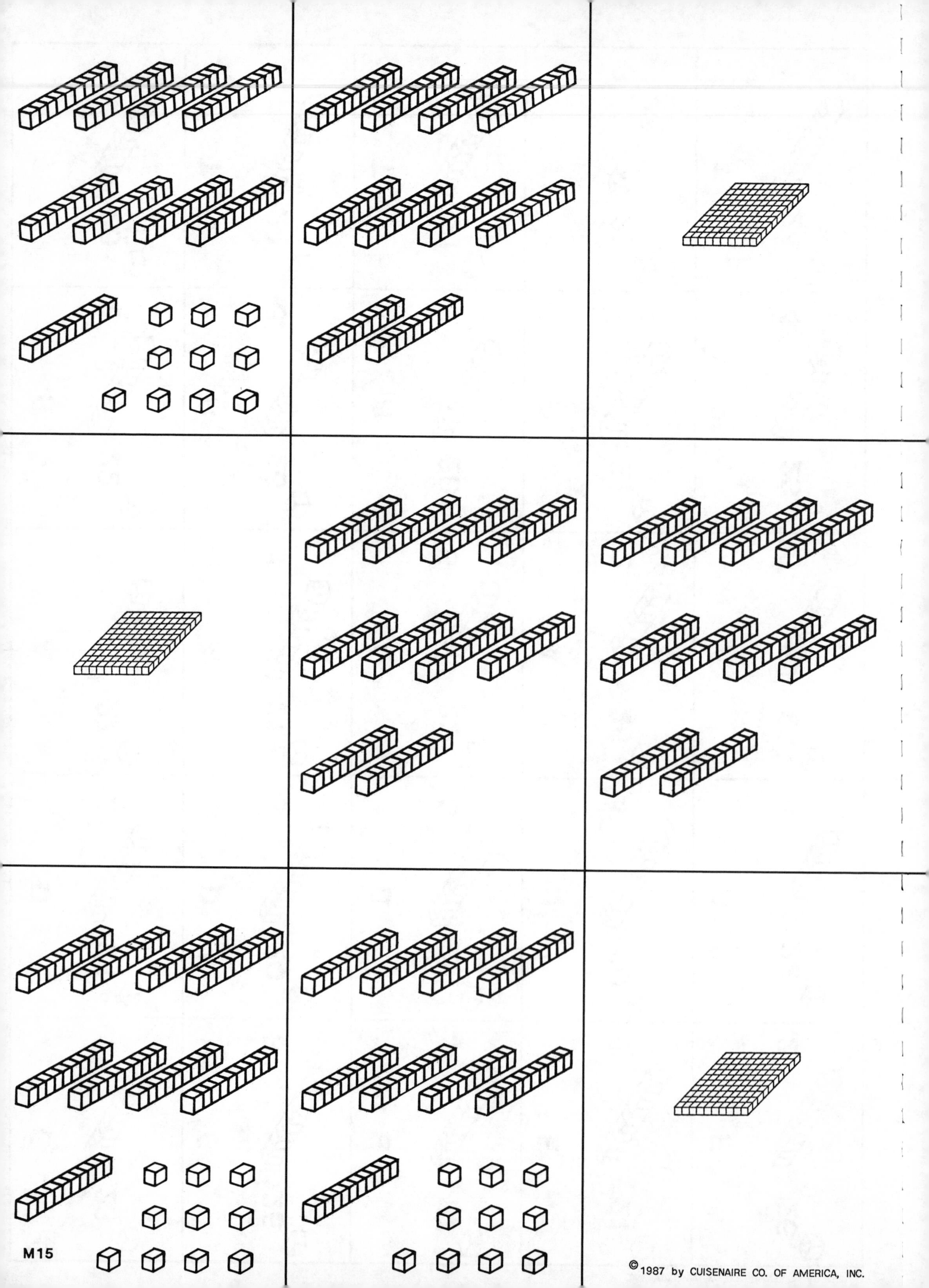
M15

thousands	hundreds

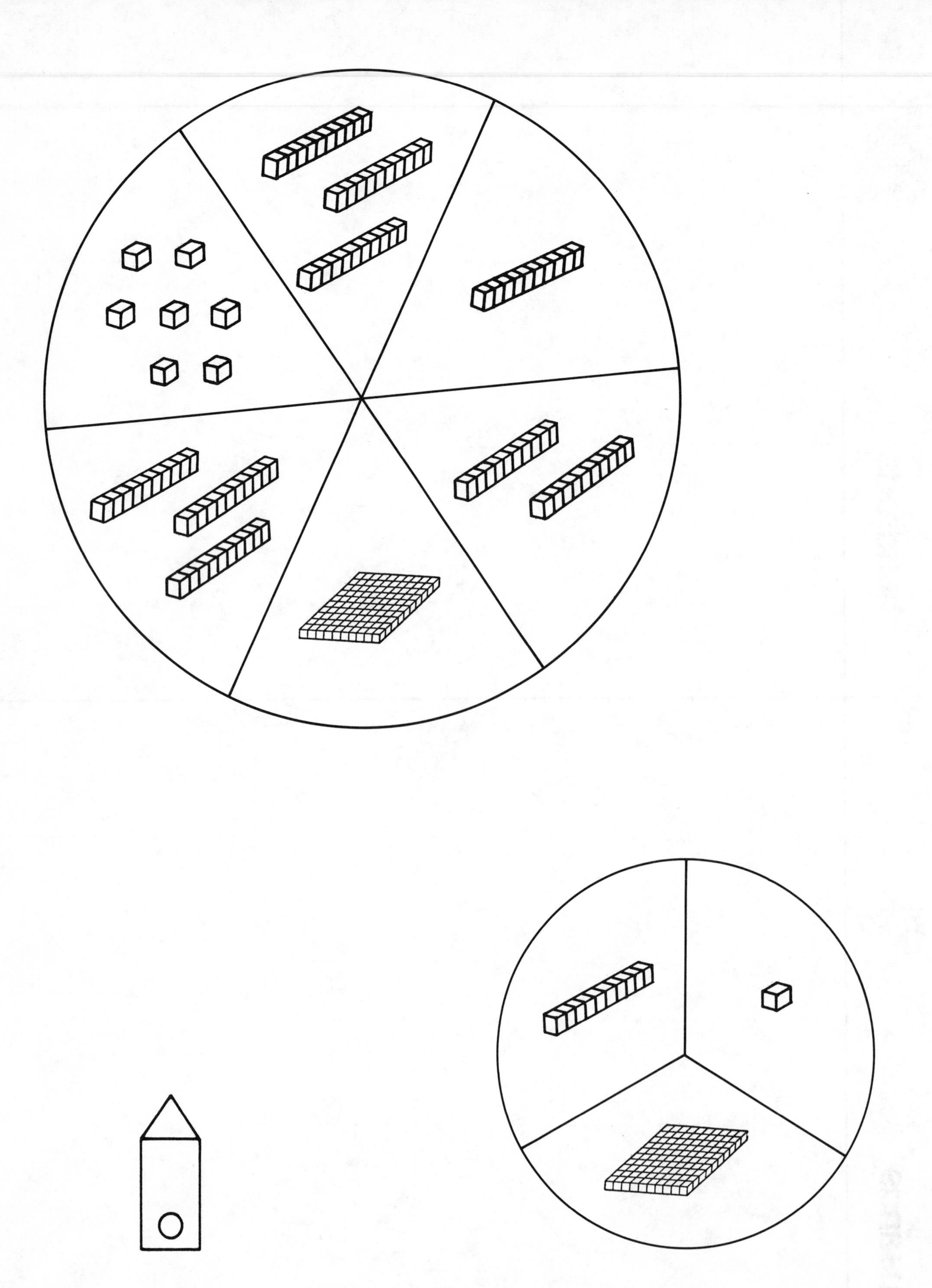

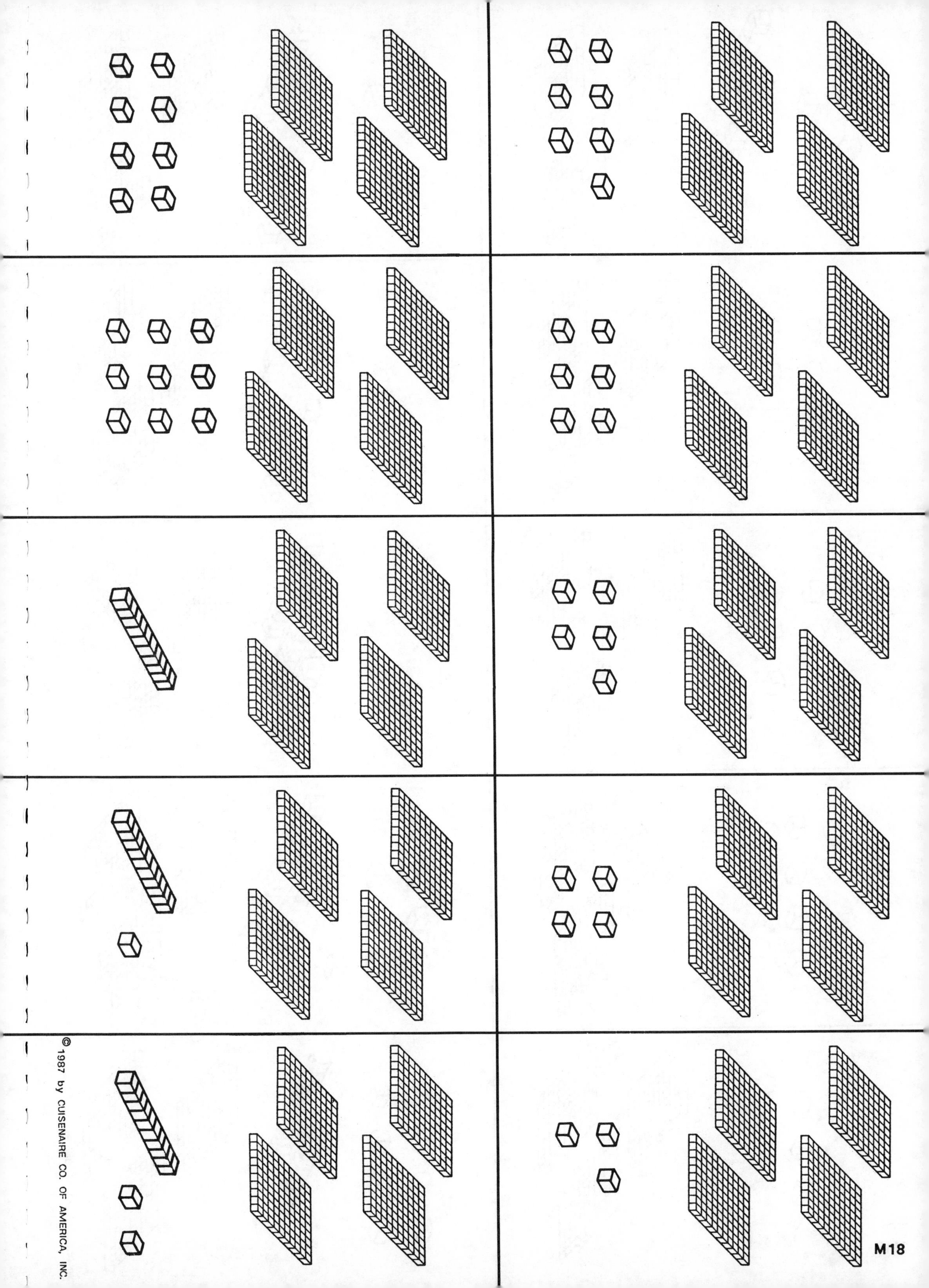
M18

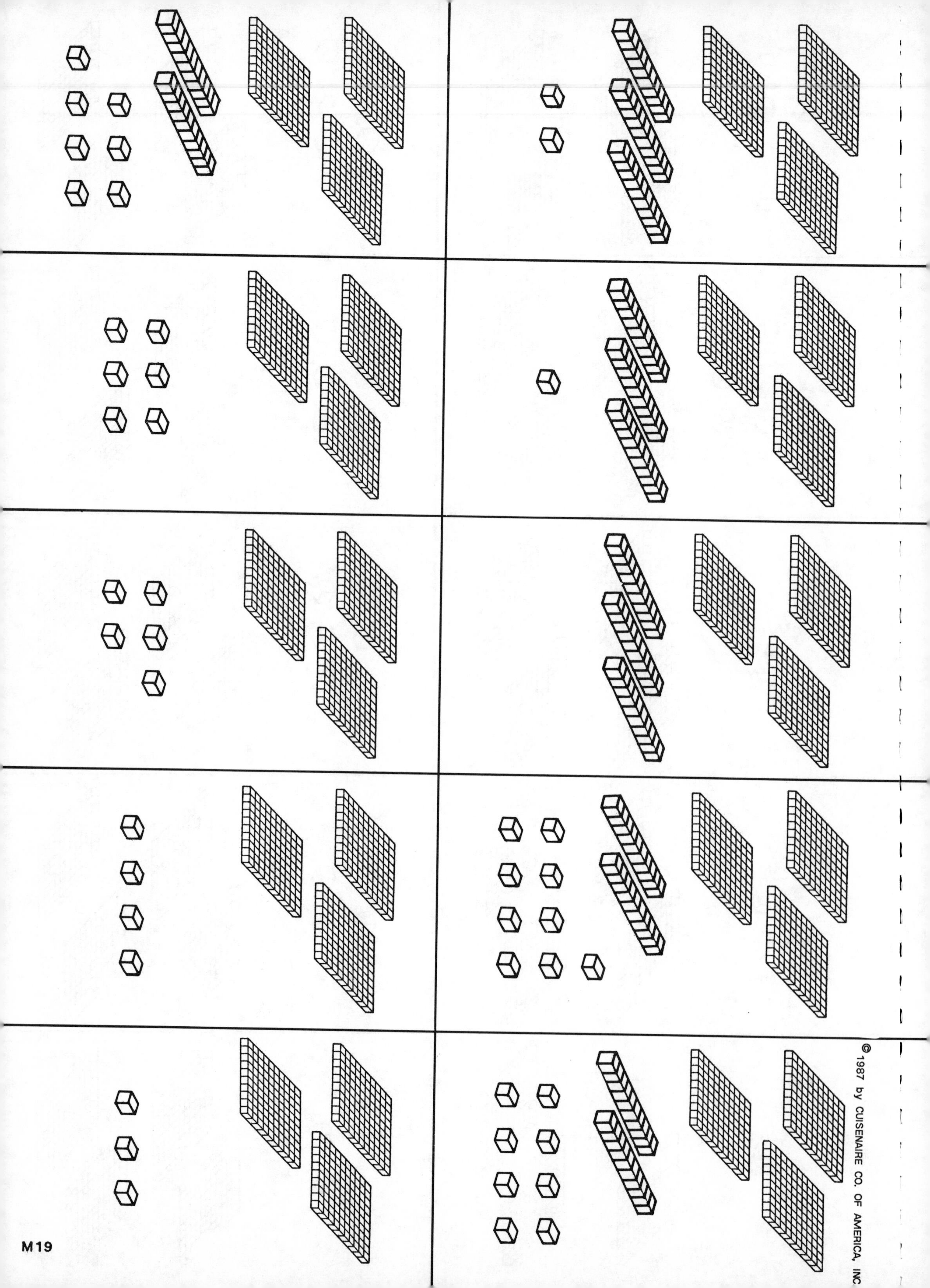

M19

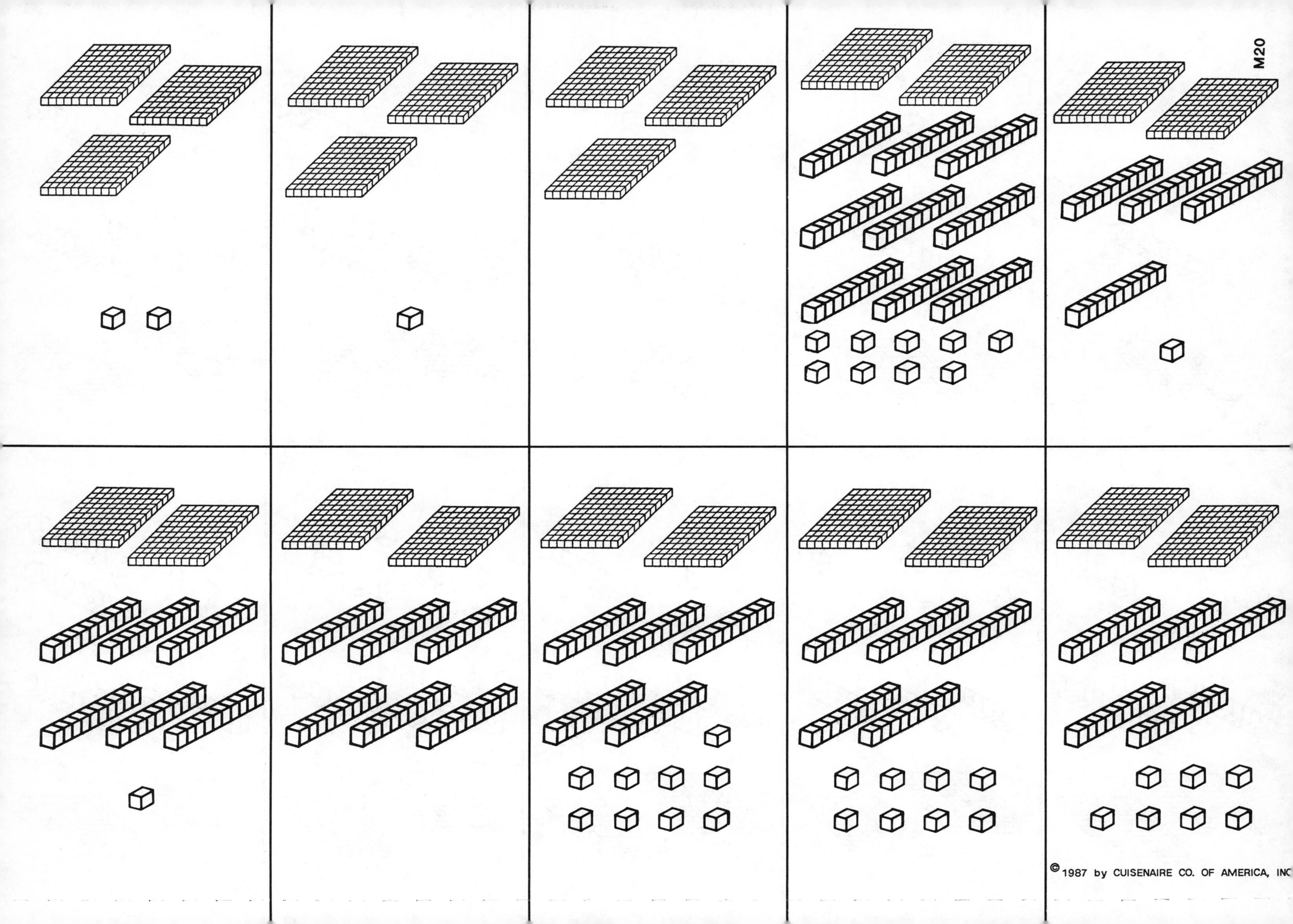

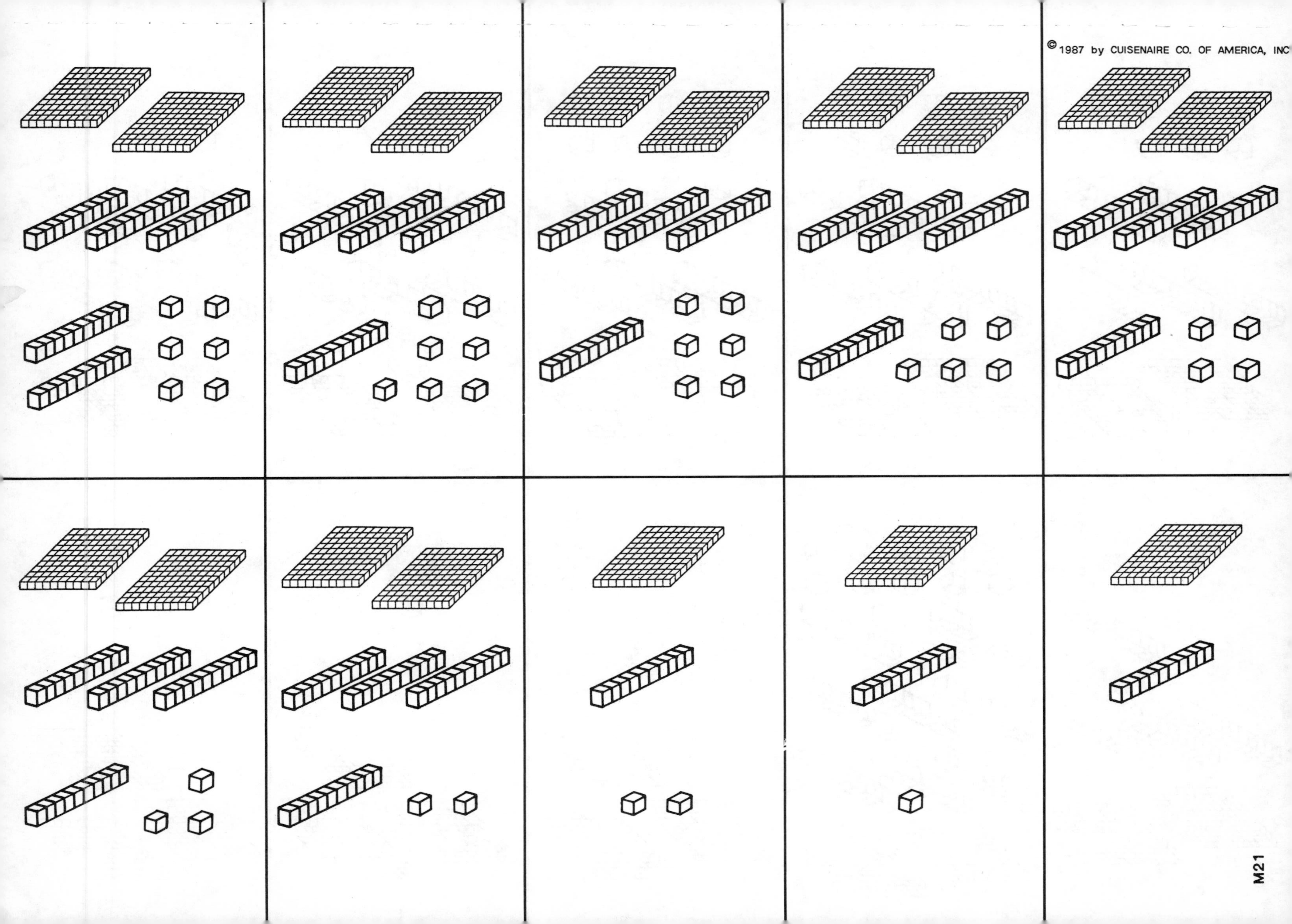
M21

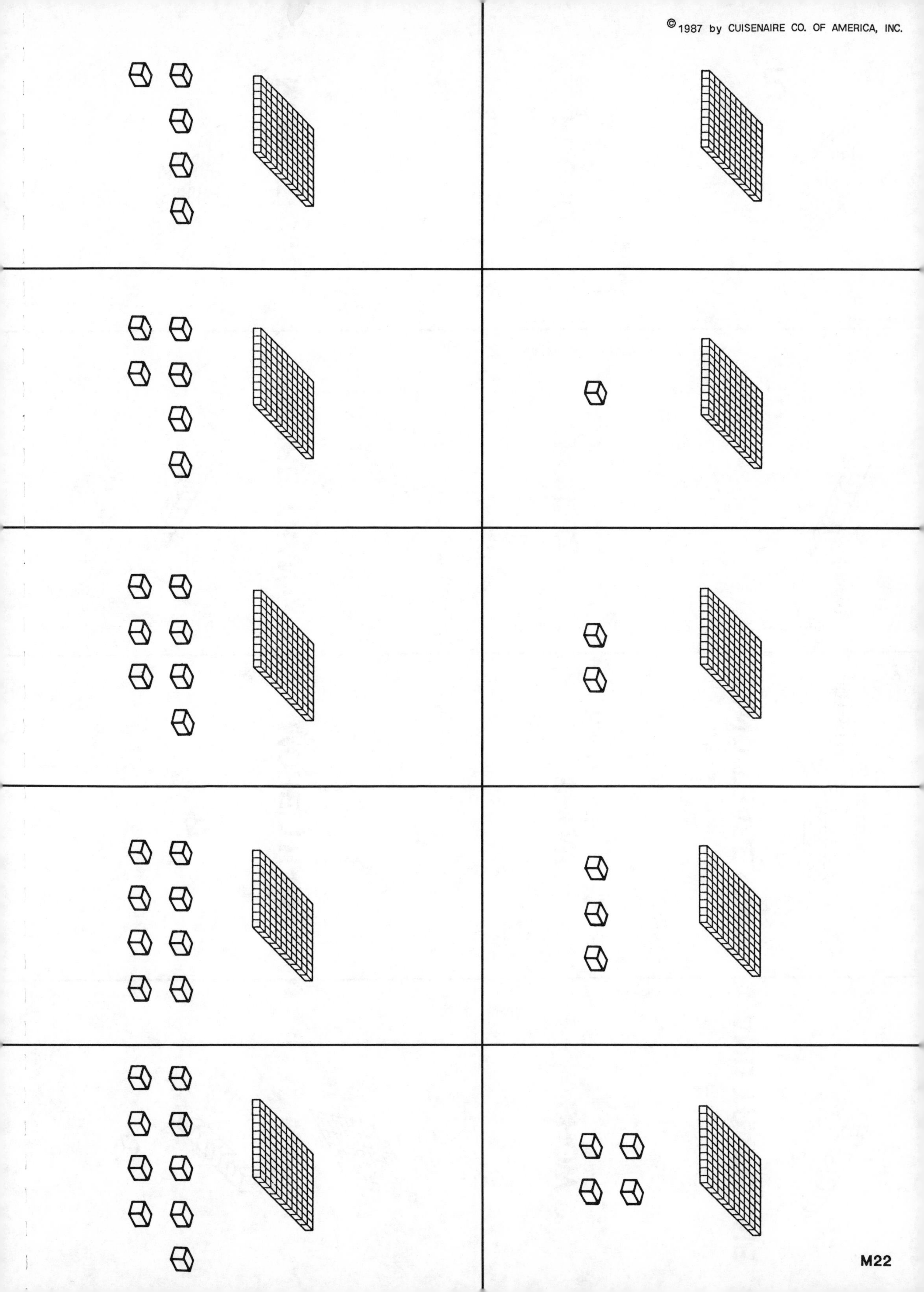

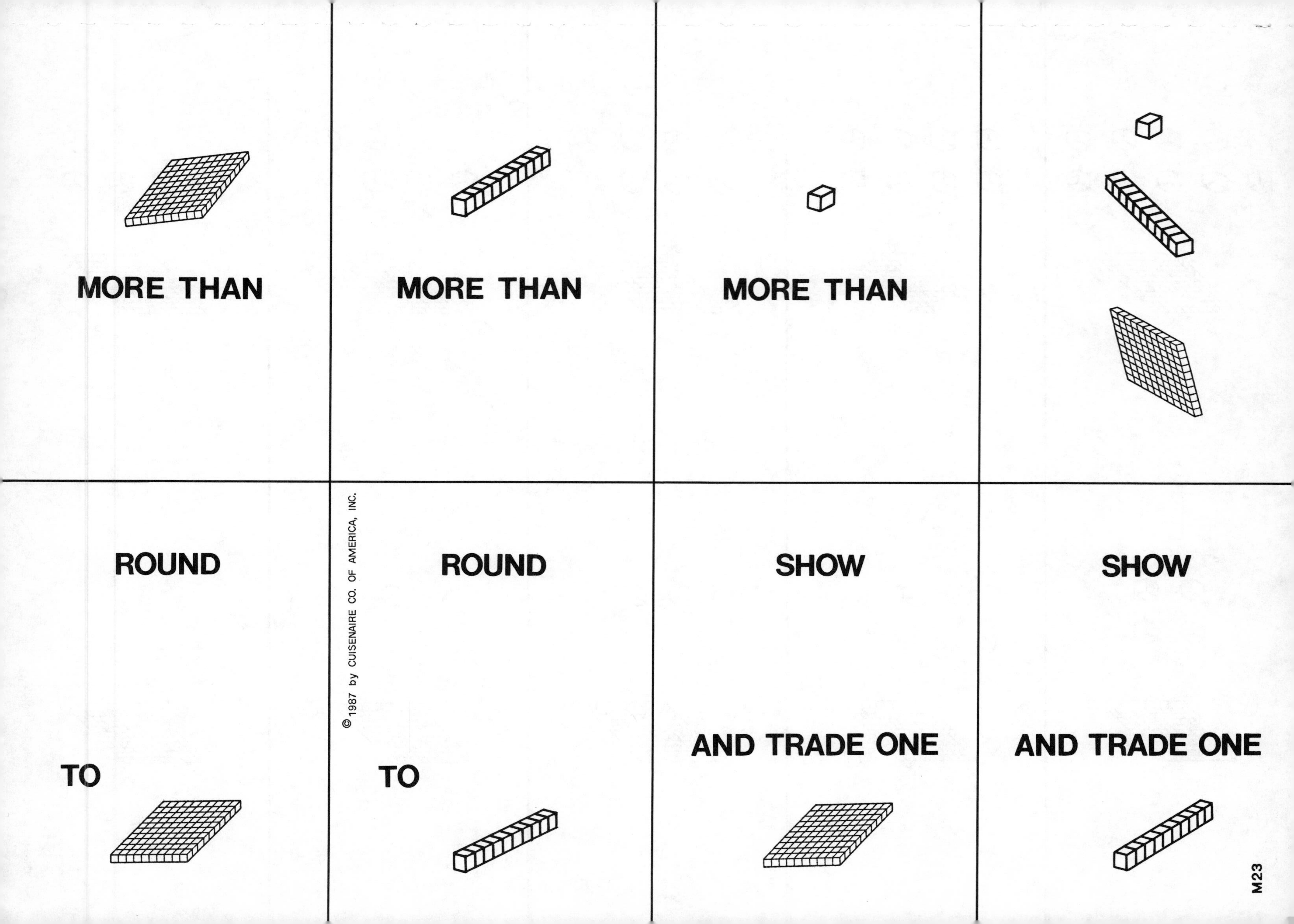
MORE THAN
MORE THAN
MORE THAN
ROUND
TO
ROUND
TO
SHOW
AND TRADE ONE
SHOW
AND TRADE ONE

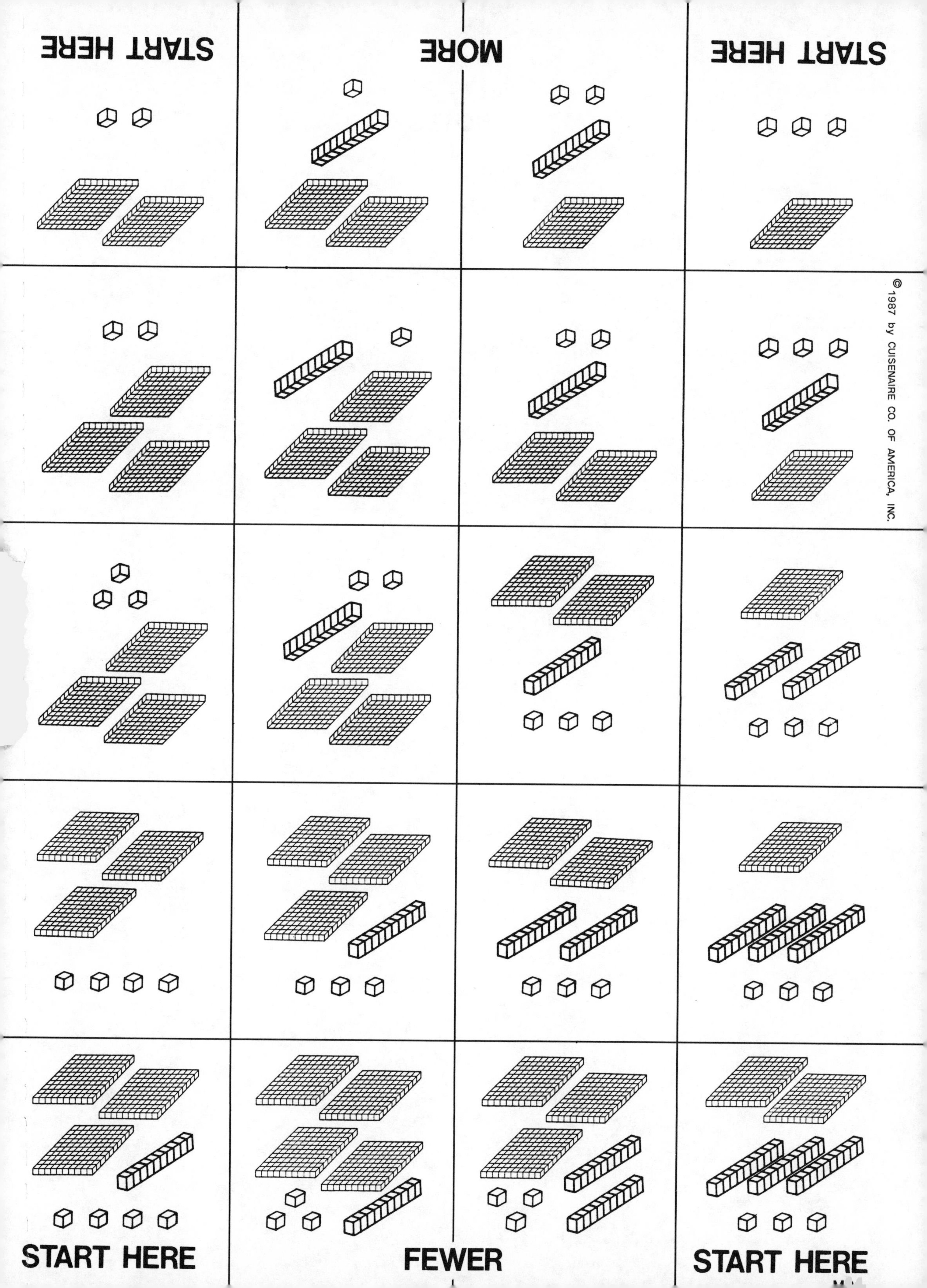
START HERE
MORE
START HERE
START HERE
FEWER
START HERE

NOTES

NOTES

NOTES